KEEP LOOKING UP

Your Guide to the Powerful
Healing of Birdwatching

TAMMAH WATTS

HAY HOUSE

Carlsbad, California • New York City
London • Sydney • New Delhi

Published in the United Kingdom by:
Hay House UK Ltd, The Sixth Floor, Watson House,
54 Baker Street, London W1U 7BU
Tel: +44 (0)20 3927 7290; Fax: +44 (0)20 3927 7291; www.hayhouse.co.uk

Published in the United States of America by:
Hay House Inc., PO Box 5100, Carlsbad, CA 92018-5100
Tel: (1) 760 431 7695 or (800) 654 5126
Fax: (1) 760 431 6948 or (800) 650 5115; www.hayhouse.com

Published in Australia by:
Hay House Australia Ltd, 18/36 Ralph St, Alexandria NSW 2015
Tel: (61) 2 9669 4299; Fax: (61) 2 9669 4144; www.hayhouse.com.au

Published in India by:
Hay House Publishers India, Muskaan Complex, Plot No.3, B-2,
Vasant Kunj, New Delhi 110 070
Tel: (91) 11 4176 1620; Fax: (91) 11 4176 1630; www.hayhouse.co.in

Cover design: Micah Kandros
Interior design: Karim J. García
Interior illustrations: Images used under license from Shutterstock.com and iStock.com/ilbusca, pg. 59

A catalogue record for this book is available from the British Library

Tradepaper ISBN: 978-1-78817-631-6
E-book ISBN: 978-1-4019-6335-4
Audiobook ISBN: 978-1-4019-6472-6

MIX
Paper from responsible sources
FSC
www.fsc.org
FSC® C013056

Printed and bound in Great Britain by
TJ Books Limited, Padstow, Cornwall

To my mother, Fran—the rarest bird of all.
Mom, I love you.

CONTENTS

WHY BIRDS?

For centuries we have belittled the bird and relegated it to a station in life of near unseen existence, by many, in servitude to others and treated as pestilence. Yet these mighty beings who sport feathers and take flight—as did their ancestors, the dinosaurs—have remained in our lives, helping and nourishing our bodies and souls. Ironically, we have caged and studied birds to inform us about ourselves.

Birds continue to defy our limited understandings and expectations of them. They can elude gravity, travel equal to the distance around the world, communicate in their tongue and ours, and demonstrate calculated decision-making, self-awareness, consciousness,[1] and the ability to see to great depths and distances. It is now known that despite differences of anatomical and structural arrangement of their forebrains, both birds and mammals possess[2] brain cell neurons and therefore similar cognitive brain function.

In fact, birds have more brain cell neurons than do primates. The more intelligent species, including parrots and many songbirds who engage in play and fashion tools, have twice as many cell neurons.[3] And among corvids such as crows, ravens, and jays, they possess even more. Birds' brains allow them to learn and vocalize their songs, and to go to sleep while miraculously keeping half the brain awake and alert to monitor potential predation.[4] The brain of the bird is truly phenomenal—bird brain indeed!

Despite the change of seasons and unforeseen man-made and naturally occurring events, birds miraculously recall where they have been and where it is that they need to go in order to not just survive, but thrive, from generation to generation.

Nature is steadfast, true, and regenerative. In its cycles are opportunities to begin again—renewed, informed, and enlightened. We also have the same opportunity.

The pandemic caused most of us to pause and take notice, to slow down or completely stop. Nature and wildlife benefited.

Nature benefited from the slowing of traffic; the silencing of saws, hammers, and engines; and the considerable clearing of the air, as evidenced by satellite photos. Animals began to emerge from the borders of existence to flourish and frolic in empty streets and uncluttered passageways. Sea life came closer to shores and into waterways absent of motoring boats and nets.

Miraculously, birds and their songs became prolific beacons around the world, pronouncing the will for us all to remain optimistic. Have you taken notice?

This book is and isn't about birds—it's about life being curvy, twisty, and imperfectly perfect. It's about connection and joy, despair and healing. It's about you, and me, along with all the rest of us throughout the world. And most certainly yes, it is about the powerful healing that birds can have in your life as well.

The practice of birdwatching helped me with my struggles from years of debilitating pain following a surgical procedure and subsequent depression that had left me in secret despair and feeling hopeless for a cure. In my darkest of hours, the connection I formed with a little yellow bird just outside my kitchen window splashed sunshine all over my soul. I rediscovered hope, love, support, and life beyond the confines of my four walls. At first glance I'd say that it was a sheer accident that I discovered birds and their powerful connection, but as Louise Hay said, there are no accidents. My mother, Fran, who is most certainly aligned with her, says and believes the same.

It is my hope that you will find fortification and resonance from the challenges, triumphs, therapeutic lessons, and strategies I share with you and that you will feel compelled to embrace our feathered friends that are all around.

No matter where you are, what you look like, and what you may or may not have or do, you can create sacred space and connection with birds and heal. Birds transcend language, place, supposed stations in life, and even time. They are what we all have in common, regardless of where we live. Birds reflect the health of our world.

So, to answer the question many have asked me, "Why do you love birds so much?", this is why. Birds are awesome!

They remain a constant in my life, and my life's journey is so much richer because of their grace. To honor their pure and miraculous elegance is to honor ourselves the same, just as our ancestors have done since time immemorial.

Peace + Birds,
Tammah

INTRODUCTION

First, I want to extend a warm welcome to you.

Just as nature is flexible and adaptive all at the same time, so too this book is a narrative memoir, an easy-to-use guide to learn how to practice birding, and a transformational self-help guide that includes simple therapeutic exercises to deepen your understanding, appreciation, and practice in context.

You are encouraged to do what you feel comfortable doing within your capacity, though I encourage you to consider exploring new horizons along the way whenever possible. There are no specified timelines, and you are invited to personalize the process to best suit your needs.

This book is constructed so that you acquire information in a cumulative fashion, beginning with birding from the inside, where you live, to eventual travels to distant new lands and beyond for lifelong pursuits. It is my hope that you gradually develop the skills to participate in bird outings near your home and as far as you wish to explore. You can take the guidebook along with you for easy reference.

Each chapter begins with a story—my personal story—in which I share experiences and reflections from my life with birds and birdwatching that have profoundly influenced me and have contributed to who I am. Zora Neale Hurston's well-known quote, "There is no greater agony than bearing an untold story inside you,"[1] heralds what I believe to be the very essence of who you and I are—our moment-to-moment lived and witnessed experiences over time. I was rather surprised by which stories came forth while writing this book, and I consider them my gift to you to help you know that your life and all that you have lived are worthy of your recognition and your telling them, free from stigma's imprisonment.

We are our stories, and our stories often give testament to our powerful connection with the more-than-human world when we grant ourselves permission to pause and take notice.

I also provide a Starter Tool Kit so that you have a quick, at-a-glance guide for the tools you'll need related to each chapter's theme. Then I offer Exercise Prompts to help deepen your personal growth process. They can be done as often as you desire. And last, a Reflection Pond section offers you an opportunity for introspection about yourself and your connection with birds and birding experiences.

I encourage you to use your own journal or sketchpad to write, sketch, or draw to your heart's content. Use pen, pencil, markers, adhesive for pictures and found items, paint, or whatever medium best expresses your thoughts or observations. Some may wish to chronicle the birds they see, while others may focus on wild and plant life. Use your creativity and enjoy!

At the end of the book, you will find a list of global resources to continue your journey and relationship with birds.

A special note:

Throughout the book I make reference to birdwatching and birding. Technically and historically speaking, the two have meant different kinds of practices to watch birds. *Birdwatching* was and is still frequently used to refer to the casual, amateur, and enthusiast level of engagement, whereas *birding* traditionally refers to the hobby/sport of actively seeking out species of birds and possessing extensive knowledge of ornithology.

There is an upswell to eradicate this notion of differentiation and to refer to the practice of watching and spending time with birds as *birding*. This term is inclusive and for the purposes of this book, I elected to most often use the terms interchangeably to include all skill levels, abilities, manner of engagement, and level of interest.

I also make reference to positioning oneself in various ways throughout the book, including sitting, standing, walking, looking, listening, and seeing, among others, and wish to emphasize that you can adapt how you do them based on your ability, preference, and need.

CHAPTER ONE

NEST

Birdwatching at Home

Wherever I fly from my own dear nest,
I always come back, for home is the best.

– Maud Lindsay

My first vivid memories about birds are as a young Black girl growing up in my family's nest in San Diego, California, during the early sixties. Both my parents, Jim and Fran, heralded from Black family traditions rooted in the South with a distinctive defiance for ingesting prescribed stereotypes and their confining expectations. My little brother James and I were "The Chosen Ones," both having been adopted, me at two days old and, five years later, James, Jr. at three months of age. We lived in a modest

middle-class neighborhood nestled in a pocket of the city that still had dirt roads that dipped and bumped the cars to and from their homes. We took pride in the fact that smooth asphalt was all around on our street except if you turned right at the bottom of the big, dangerous hill.

Our house sat at the end of a row of newer, modest-sized homes situated on half-acre lots. From our large-paned living room window, we could see the "Old House Across the Street," whose door opened out to a raised mound of earth confidently impersonating a porch.

Withered white paint curls stuck out from shutters, posts, and door. Weeds and an overgrown pepper tree jutted up from the backyard amid what was otherwise barren hard dirt mixed with some gravel. No one had lived there for a very long time according to my mother, and that got her to humming. Whenever my dad was away in service to our country courtesy of the U.S. Navy, my mother's humming took on a life of its own. It permeated all hours, off and on throughout the day as she tinkered in the kitchen or sat on the blue sofa and tweezed her errant eyebrow and chin hairs, and into the late night after the talk radio had gone silent.

I hated it. Even as a child I knew the humming had a nervous, pent-up quality to it that wasn't good. For months she hummed, tweezed, voraciously read book after book, and fretted aloud about the Old House Across the Street.

And then one day out of the blue she announced, "Come on, kids. We're going across to take a look at the Old House Across the Street."

I was terrified. I knew this wasn't supposed to happen; we didn't have our neighbor's permission, and what if something did happen in that scary old house? But the three of us crossed over and ventured into the weeds and dirt. Fran ushered us to peer in from one window to the next while my brother, little Jimmy-Jim, ran and crawled around the base of the house's frame. We dared to explore beyond the back of the house and into the vast, open lot behind it that sloped steeply midway down.

To the right leaned a decrepit wood structure that my mother instantly fell in love with. I waited outside while she and Jimmy-Jim went in. A rooster's crow startled me; it was so loud that I was certain it was somewhere in the dense thicket.

From beyond the fence, a high-pitched old lady's voice crackled out: "Hello? Hello over there! Can I help you? What are you doing over there?"

Fran and Jimmy-Jim reappeared and walked over toward the voice. I slowly and ever so carefully followed and saw them talking with an elderly white lady dressed like someone I'd read about in *Little House on the Prairie*. Fran beckoned me to join them. Chatter led to formal salutations and eventually an invitation from Mrs. Poulson to come around the side, through the gate, and into an equally immense yard filled with huge old trees and an assortment of wildflower patches scattered among piles of rusted drum barrels and twisted branches.

Mrs. Poulson introduced us to her one rooster and many hens. The chickens were a warm tawny brown, creamy white, and jet black with flecks and spots, and all of them scratched the earth for no apparent reason.

This was the first time that birds were brought into my life.

We made several more visits over to Mrs. Poulson's, which were highlighted by my mother's lovely smile and full, melodic laugh that could put anyone at ease. Later I learned that our visits had been strategic, as she would eventually hand Mrs. Poulson an envelope filled with cash and a bundle of signed papers.

One day, following our now cursory good-byes with Mrs. Poulson, my mother walked Jimmy-Jim and me directly next door to the Old House Across the Street, inserted a key, and pulled open the resistant door, proclaiming, "This is our new house, kids! This is going to be your farm!"

And following a dizzying whirlwind of house cleaning and clearing of the overgrowth on the property, along with Fran's rebellious commission to have the Old House painted completely charcoal black with hot-fuchsia trim, we had our version of our newest nest, a farm.

The old shack was nailed erect and soon provided shelter for our Shetland pony named Timmy; five rooster chicks, four that Fran named after the faithful disciples Matthew, Mark, Luke, and John, and one rooster chick named Rocky; and my own personal pet yellow duckling. My relationship with Fluffy Duck proved pivotal for me, even at just seven years of age.

As my mother hummed away while wallpapering and decorating the farmhouse that we also lived in for an extended period of time, I developed a close and tender relationship with Fluffy Duck. At first, she was allowed to stay and sleep in an infant's playpen in the kitchen where she could be warmed by the oven. But as her feathers changed from a rich golden yellow to pure white and she increased in size, my mother declared it was time for her to be outside. Fran made Fluffy Duck a modified coop under the newly constructed wooden deck of the bachelor's studio she had built on Sears, Roebuck & Co. revolving credit.

When Dad left on yet another six-month deployment, Fran's humming got persistently louder and all-encompassing, with particular emphasis sung aloud to certain lines of popular songs. The hum would become a bellow about a woman's right to walk in patent leather boots or to toss flowers from up on a bridge into muddy water, on repeat. During that time, I clung to Fluffy Duck for sanity. I carried her around in my arms for hours at a time while she slept with her bill nestled in the warmth of my underarm. She understood and loved me, and me, her.

Fran would sometimes organize the neighborhood kids to come over and participate in garden planting and harvesting parties and spend time with our animals, and that was when we all flourished. But the calm didn't last. Eventually, some of the older boys that lived next door to our Yellow House would ask where our roosters had gone. I wondered too.

I later learned that Fran, running short on money and having exhausted leveraged credit, was anxious that Jim would find out, so she asked Mrs. Poulson to kill the roosters one by one as symbolic sacrificial lambs in the name of Jesus Christ in hopes that a miracle would somehow manifest in our favor. That included us kids unknowingly eating the disciples she'd fried with tears. All of the roosters except for Rocky, whose given name was Judas, so named because he bullied and attacked the other chickens and us humans without mercy. It didn't help that everyone in our neighborhood, kids and adults alike, were afraid of him. Rocky guarded the property ferociously as though it belonged to only him and attacked the heads and necks of all perceived trespassers. The only person he left alone was Fran. To quell alarm and the charge of blasphemy from those who visited, however, she used the pseudonym Rocky, after Rocky Marciano. As a child, I often felt as though I was aiding and abetting in a terrible secret.

Eventually, Rocky hopped the fence and joined Mrs. Poulson's flock. Fran surmised he was lonely being the only rooster left living with us since she couldn't justify eating so much evil. We eventually stopped checking on him, especially after he stopped looking our way, but he was not forgotten. To this day, despite my love and admiration for birds, Rocky has left in me an indelible fear that whispers when a bird flaps overhead.

Yet equal to the terror I had of Rocky was my immense love and spiritual connection to my Fluffy Duck.

While I continued to appreciate birds and wildlife as an adult, I lost the deeper connection I once knew and cherished as a young girl. It had gotten lost amid life's many obligations and commitments: relationships, children, education, and career. All that came with tireless efforts for balanced perfection, and somewhere along the way, nature and wildlife became imperceptible to me.

My career and personal life bloomed until my required laparoscopic hysterectomy, an ordinarily routine surgery with a relatively uneventful recovery for most women. But for me, the scorching pain in my veins from the IV at the hospital worsened during my convalescence at home, and I was ultimately placed on unexpected disability leave from work. I had been confined to my nest, a place that was unfamiliar, since I had barely spent time in it due to work and an active life.

I can still recall how full of despair I was that morning standing at the kitchen sink, seemingly trapped inside my nest. What day *was it*? I had not been able to answer that question for several days, and the reality frightened me. As my entire body trembled from pain, I realized that I stood career-less, disfigured, broke, and disabled. I was ashamed, angry, indignant, and completely alone. Or so I thought.

At that moment, I caught a glimpse of warm butterscotch yellow flutter up momentarily on a tipa tree branch that reached down toward the kitchen window. I stretched to look through the window and into the tree's delicate oval-shaped leaves and tiny clusters of yellow blossoms to be certain it had in fact been a bird. And there it still was, flitting from one branch's cluster to the next, casting its beams of sunshine all over my soul. The little yellow bird seemed to be just as curious about me as it gazed down with a tilted head, our eyes meeting ever so briefly before it flew away.

In that instant, I became reacquainted with the memory of hopes and joys and in knowing that gifts from the Universe present themselves in one's darkest hours when light can truly be seen and appreciated.

In the days that followed my divine chance encounter, I began to look for my little yellow friend every time I went into the kitchen. Stretching over the sink in order to look high up among the branches became a regular part of my day, and the most routine I'd had in years. I began to see more birds sitting, flitting about, singing, and eating from the tree leaves. They were very small and greenish yellow, medium-sized and brown with dark spots, and the drabbest of browns with crests of orangey red.

Over time I learned the patterns of the birds that lived in and around my yard. I began to set my watch to 10 each morning, the exact time the tiny little bushtits would flock to the tipa tree, announcing their arrival with signature high-pitched squeaks. Not much bigger than a hummingbird, their round, brown bodies urgently scurried and skipped along the tree's branches, devouring bugs that I never could see.

The more I watched the birds in the front and backyards, the more I felt connected to life's smallest wonders. My depression began to improve and gave way to hopefulness and curiosity about life and living beyond the four walls of my home. Slowly I began to notice that the time spent watching for birds was time *not* spent focusing on my pain and isolation, and I was eager to increase that time each day.

My backyard became my refuge and salvation. I spent more time outdoors, reconnecting with the simplicity of life that I had lost. It became a blessing to be back in my nest, for it offered me the protection, shelter, and gifts I would not have otherwise seen. My love of being in nature, which had lain dormant from grief over my once-active lifestyle, was reawakened.

Even the process of time and the changing seasons for the birds could not make my resolution waver; if anything, it only strengthened my care for my feathered friends. Winter, even for Southern California, can be challenging for wildlife. I had developed bonds with some of the regular residents: boisterous house finches, demure mourning doves, and true to name, singing song sparrows. I worried about them making it through to spring. I bought a book and a bird feeder after looking online to figure out what was what. On days that I was unable to move, I persisted to still find a window and spend a few minutes watching and listening for the birds, who had already proven to be doses of powerful healing.

If you've had the blessing to watch a bird build its nest, then you know that much time and care is given to the construction. Different species of birds build different nests and use different items found in their environment, including natural and man-made materials such as branches, roots and twigs, dryer lint, spiderwebs, vines, animal fur, and string. Some are the size and shape of a tiny teacup, an elongated hanging bag, a molded mud pot, or a large, wide basket crisscrossing between large tree branches, while others are carved-out burrows in trees or earth formations. Bird nests can be found in trees, under eaves and on ledges of buildings, in and under brush and thicket, and on the ground.

Regardless of their differences, all birds find mates to build nests with for the same purpose of providing a nurturing environment for their eggs and their young to flourish. The care for their young is most often done by females, although some species share in the raising of their nestlings or are cared for by only the male. When the fledglings are strong and ready to leave the nest, they must survive a tenuous first year fraught with predation and frequent starvation.

Much like our feathered counterparts, the essence of who we are is in large part predicated on how and where we live and what we hold dear. Our nests, or homes, provide us with shelter, personal comfort, nurturance, and safety, not only when we are children, but also as adults.

We depart our nests at a tender age to find our way in the world. But as fully grown birds will eventually return in future years to build nests of their own, we also return home in times of need, for connection, or when having troubles. We carry home in the depths of our soul and many can call on it like a long-passed ancestor to restore us, to hold us, and to assure us during tender, vulnerable times.

Birds craft their nests with innate devotion to their eggs in order to successfully nurture them to eventually take flight. You can do the same for yourself in your own nest by creating a safe and nurturing environment in which to not just live, but thrive.

I wholeheartedly invite you to embrace the healing power that watching birds can have in your life. Nature by design is elemental, earthbound, and free of judgment. As such, you can engage in birdwatching at home completely on your own terms, connecting with the outside world even while remaining indoors or just a few steps beyond. If you cannot sit up, then remain lying down, and if you are unable to see, then listen. Birdwatching at home allows you to feel safe in familiar surroundings and can reduce symptoms associated with stress, anxiety, pain, and depression.

Birdwatching expands and extends your physical living space, creates opportunities for you to connect meaningfully to life and living, and honors nature's inherent ability to heal what ails us. Watching birds at home returns us all to our symbolic nests and allows us to nurture and tend to our inner child just as we do for the birds who come to visit. Nature is healing indeed.

I encourage you to venture outside near your home and gain confidence birdwatching and continue to cultivate resilience to persevere in other areas of your life. Perhaps the biggest gift of all is the pure joy and fun that comes with spending time with our feathered friends from your very own nest.

FROM INSIDE LOOKING OUT

Take some time and watch for birds from inside where you live. Home, our nest, is where the heart is, and it is unique to each of us. Many of us live in houses, apartments, mobile homes, condos, trailers, and cottages. Some are one story, while others are two floors or many more. Others live in care facilities, hospitals, or unique shared living spaces. No matter what you call home and where you live, if you have a window, doorway, porch, deck, stoop, yard, or similar space, there is always a wonderful opportunity to watch for birds.

If you spend a large portion of your days at work or school, and may even feel as if it is your home away from home, I encourage you to take a few minutes out of each busy day and devote yourself to birdwatching to reduce tension and stress and increase your ability to focus. After all, in the case of migrating birds visiting your area, it is their home away from home too.

Choose a window or doorway that offers you the best vantage point to see outside. Ideally, choose more than one, if you can. Views of an entire yard or area will offer you the best chance for seeing different birds throughout the day.

Sit or stand as still as possible while watching. Birds are hyper aware of their surroundings and are always on the lookout for predators. Sudden movements will frighten them and cause them to retreat or fly away. That said, it's also important for you to be comfortable while birdwatching so that you have the best experience possible and will want to engage again and again.

Listen first. Allow the sounds you hear, such as chirping, tweets, and calls, to guide you in their general direction. Over time, you may recognize a bird's song or call before actually seeing it.

Use your eyes to scan the area you want to watch. Look for movement in trees, bushes, and on the ground. Once you see motion, focus your eyes on that spot and wait for more. Eventually, you will see the birds making the commotion, and you can then track them as they flit and fly.

HOW TO IDENTIFY THE BIRDS YOU SEE

Size and Shape: Figure out the size of the bird relative to the other birds you are viewing or know for comparison, for example, smaller than a dove, larger than a sparrow. Determine the shape of the bird's body, head, and tail and try to see its silhouette.

Behavior: Determine what the bird is doing. Is it scratching and foraging for food? Feeding its young? Sitting still? Is the bird alone or part of a group? Take note of how the bird flies.

Color: When we think about birds, most often their feathers come to mind. The magnificent array of colors is truly fascinating to see. At different times of the year, however, many birds' plumage changes, and that can make identifying them trickier. Most often, males display the more vibrant colors, while the females typically bear muted colors.

Habitat: Where the bird chooses to be is an important clue to the species. Birds tend to remain in their respective habitats, and this can help us to identify them accordingly.

(Cornell Lab of Ornithology has developed a free app called Merlin Bird ID to assist in the identification of birds.)

Optional Considerations

Binoculars (and scopes) are a great tool. Using binoculars is pretty essential in observing the fine features and details of birds, especially from long distances. Start out with a pair you may already have, or purchase a basic, functional pair. As you adapt to using them on a regular basis, consider investing in some with higher-quality and advanced features, and consider scopes as your preferences dictate and finances permit.

Jot down some notes about the birds you observe. This is especially important in the beginning when you are learning about different species. Writing down key features, behaviors, and where in the environment you see certain birds can be very helpful when trying to identify its species.

Sketch the bird. A quick sketch of a bird's key features is another way to record information to reference later when trying to identify what you saw. Don't feel intimidated by this approach, as there is no requirement for you to be an artist; you'll be surprised at what you remember and how much enjoyment this process of putting pencil to paper offers.

Use a field guide. A rewarding part of birdwatching can be identifying the birds you observe. Gradually exploring field guides in order to determine a species will reinforce your observation skills and broaden your knowledge. Field guides are abundant and are organized by region, learning preference, portability, and depth of reference information. They are available as books, guides, leaflets, and apps. (The National Audubon Society has a free app available for iPhones and Androids).

Take photos. Your phone or a camera is an additional way to record the birds for identification purposes. It can be very satisfying to spend time looking at and sharing the photos of birds you have observed, and you can use them later for reference.

Keep a birdwatching log. Writing down what species of birds you observe, the time of day, and where in your environment you saw them can provide a record of sightings over time and serve as a reference in future seasons to understand the birds' patterns and visitation cycles. The log can also be very therapeutic, as it chronicles your participation, activity, and bird-sighting accomplishments.

GOING OUTSIDE AT HOME

Go outside where you live. Going outside is a huge step toward your personal journey of healing. Birdwatching in your yard, on your patio or balcony, or from your porch or stoop physically immerses you in nature. For some of you, it may be the first time that you have ventured out in a very long time. Be kind to yourself and to your process. Do what you can and for a time period that feels reasonable. Gradually, you may develop suitable comfort and build stamina to spend additional time outdoors.

Wear bird-friendly attire. Preferably, wear muted colors while outside in order to blend in with the surrounding environment. Birds are sensitive to bright colors and perceive white as a sign of danger from a potential predator. This said, most importantly, wear clothing that is comfortable.

Move slowly and with intention. Sudden movements startle birds and cause them to fly away. I have found that some species of birds who take up year-round residence in the yard can acclimate somewhat to my presence while I am outside so long as I adhere to bird-friendly and bird-mindful practices.

Incorporate the guidelines you use for birdwatching from inside the home for use outdoors as well.

Feed the birds. A great way to encourage different species of birds to visit and live around your home is to offer them bird food on a regular basis. You may want to start with one bird feeder of good-quality seed specific to your region. Feeders come in different shapes and sizes, and different species of birds have unique dietary preferences: black-oiled sunflower seeds, thistle and nyjer seeds, millet, and suet are common varieties.

Provide the birds with water. Fresh water is essential to a bird-friendly habitat. A fountain or feature that circulates the water is ideal. If you do not have a fountain or water feature, you can also use a large pot or tray filled daily with fresh water. Strategically place it on the ground near plants for ground-dwelling birds. A flat rock can be placed inside it to steady the birds when they enter to bathe or drink.

Provide shelter. By offering the birds houses of their own, leaf litter, and covered spaces, you increase the likelihood of permanent tenants and the opportunity to observe them over long periods. It can be gratifying to see birds inhabit a dwelling and make a nest that you can buy or construct yourself.

Birdwatching expands and extends your physical living space, creates opportunities for you to connect meaningfully to life and living, and honors nature's inherent ability to heal what ails us. Watching birds at home returns us all to our symbolic nests and allows us to nurture and tend to our inner child just as we do for the birds who come to visit. Nature is healing indeed.

STARTER TOOL KIT

There is very little you need to get started birdwatching at home.

In as little as **10 minutes** you should see at least one bird and reap the mental and physical benefits from engaging in the activity.

Find a comfortable spot to **sit or stand.**

Use **binoculars** to see the bird's details and for long-distance viewing.

Use a **field guide, book, or app** to identify the birds you see.

Keep a **birdwatching log** to record your sightings and sketches.

EXERCISE PROMPTS

I am

inside outside

I am at

home work school other

This time, in order to observe the birds, I choose to:

sit stand lie down

1) Listen for birds. *While you scan the area with your eyes or ears, looking for signs of birds, take a slow, deep breath, hold it in for four seconds, exhale slowly to a count of five, and repeat.*

I hear:

What do you see?

Trees _____ Bushes _____ Plants _____ Flowers _____

Grass _____ Dirt _____ Sidewalk _____ Pavement _____

Roads _____ Rocks _____ Fence _____ Gate _____

Wall _____ Roof _____ Eaves _____ Ledge _____

Post _____ Window _____ Water _____ Wire _____

Where is the bird?

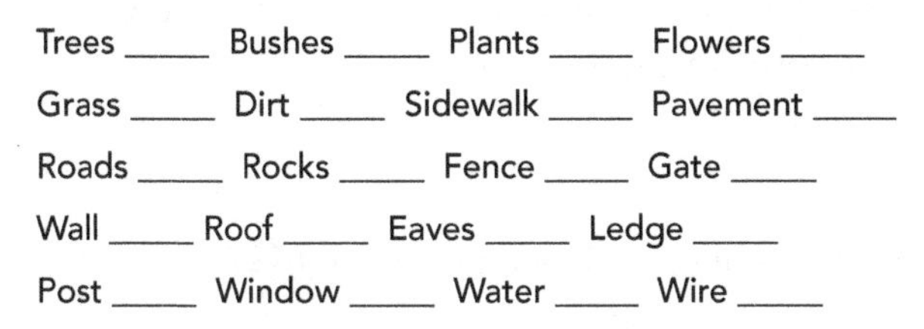

Trees _____ Bushes _____ Plants _____ Flowers _____

Grass _____ Dirt _____ Sidewalk _____ Pavement _____

Roads _____ Rocks _____ Fence _____ Gate _____

Wall _____ Roof _____ Eaves _____ Ledge _____

Post _____ Window _____ Water _____ Wire _____

2) *Take at least one full minute to observe the bird.*

What is the bird doing? (For example, sitting on a branch, walking through the grass.)

Why do you think the bird is doing that?

Describe the bird you are observing. Sketch the bird *(optional).*

- Size
- Shape
- Color(s)
- Eye color
- Leg color
- Beak color/shape

What is your impression of the bird? (For example, it seems to be searching for food or calling for its mate; is it curious, shy, familiar?)

My first childhood memory of a bird is:

When I was a child, my nest was:

And now, my nest is:

Reflection Pond

We cannot see our reflection in running water.
It is only in still water that we can see.

–Taoist Proverb

Answer the questions as though you are a crane standing at the water's edge, gazing into a still pond at your own reflection.

Before I went birdwatching today, I was:

While I was birdwatching, I felt:

And now, I feel:

CHAPTER TWO

FLOCK

Birdwatching in Your Community

Just show up, it said. The website said "all are welcome" to join the group who would survey the natural lagoons. I had researched the three Audubon chapters in the county and was drawn to this particular one because their website indicated that they had a nature center, monthly speakers, and free guided bird outings, including one for beginners like me.

I had been summoned to rise so very early in the dark morning at the insistence of my cell phone's alarm scooting across the top of my polished nightstand. I donned the outfit I had scribbled on an envelope's blank side the night before with all of my self-assigned must-dos so as

not to show up incomplete and somehow undeserving. I was uncertain of exact *real* birding attire, but I had seen enough nature shows to have an admittedly preconceived notion of what I believed birdwatchers looked like.

Jeans, an olive-green long-sleeved button-up shirt with a tan tank top peeking out just above the second button, my old hiking shoes thick with dust but now wiped uncharacteristically clean, and an old pair of brown socks I'd ferreted from a repurposed U-Haul box labeled for donation. Despite the Velcro tab being at its farthest edge, my black baseball cap sat awkwardly atop my braids. To finish, I'd decided that a touch of mascara and natural-toned lip gel would be appropriate given the special occasion.

But at some point, between prepping a thermos of hot coffee topped off with generous pours of almond milk and filling two bottles with freshly filtered water, I began to have a strong urge to just stop everything and stay at home. Self-doubt is a powerful and insidious entity and serves its master, Fear, devotedly. Especially if there's no internal process of accountability for one's oftentimes counterproductive thinking. That internal dialogue got turned way up the closer it came time for me to depart and as I literally forced-talked myself into the car.

As I drove, I could feel the anxiety building up in my body—my arms were tense and aching from my firm grip on the steering wheel, my neck had tighter-than-usual knots, and I could feel my heart beating hard against my chest. I couldn't listen to music to relax because I was relying on Google Maps for navigation. I employed deep breathing techniques—breathe in, breathe out, breathe in, breathe out—which made all the difference.

Once I reached the small-yet-accommodating parking lot, I sat in the December chill, clenching the steering wheel with white knuckles in contrast to my otherwise brown-hued skin. My binoculars—big, black, and with a hefty weight—had previously served me well from looking out through my windows and on outings near my home. They had allowed me to appreciate the striations of brown sported by the female house finch and the mourning dove's subtle iridescence shown in morning's dew. Now they allowed me to observe people hastily enter the building clad in varying shades of green, brown, tan, and denim as though a collective honoring of nature's organic existence. Various kinds of binoculars, scopes, straps, notepads, and cameras adorned their bodies like readied armor. Folks appeared to be mostly of European descent and of all ages, including a few teens and even some little ones circling about their parents excitedly.

I couldn't muster the courage to join them. Everyone seemed to know one another, and no one even came close to looking like me. I felt I had made a horrible mistake. Yet I had made it this far.

Before long, people began to exit the building and assemble in various clusters out in front of the nature center. A man dressed in the finest of birding attire exited and motioned everyone closer in to listen. Much like the children bursting with anticipation, I too yearned to hear and know and belong. My paralyzing fear surrendered to my curiosity, and I got out of the car.

Folks were introducing themselves and discussing the day's agenda. The guide, Tom, asked my name twice and warmly said welcome; and indeed, I felt very welcomed. Several people extended introductions with generous smiles, and I soon learned that many others had decided to embark on their new venture that day as well.

Tom asked if I needed a pair of binoculars, and I proudly declined, sharing my preparedness in tow around my neck. It would be some months later that I would realize that my binoculars had very limited visual power and that he had so kindly offered me an upgrade.

There's no sense in turning back now, I thought.

Tom shared a bit about the history of the annual Christmas Bird Count (CBC) and the establishment of the Audubon Society by a gathering of well-to-do women who looked forward to sporting their annual prized Parisian couture hats, embellished with exotic bird plumes: peacocks, snowy egrets, and great blue herons, among many others. Their gentlemen would go out for their highly anticipated annual Christmas "Side Hunt,"[1] in which the highest honor was bestowed upon whichever "side" shot down the most animals, birds included. Eventually, the gentility thought better of their ways, particularly in light of the drastic decline of waterfowl to the millineries' displays, and in 1896, Minna Hall and Harriet Hemenway founded what would eventually become the Massachusetts Audubon Society, and in 1900, Frank Chapman proposed what is now known as the Annual Christmas Bird Count (CBC) instead.[2]

We were encouraged to carpool with as few cars as possible, a practice encouraged by the Audubon Society to reduce intrusion into various environments. The route was centered on prime vantage points, including the parking lot of a restaurant. It was situated up above a body of water, dotted with bobbing ducks where huge white pelicans and seagulls seemed to fly in from the nearby ocean to rest there instead. Darting close by along tree limbs and bushes were little winged jewels—hummingbirds, who sported vibrant fuchsia and magenta collars about their throats. There were also common house sparrows flitting on rooftops and overgrown bougainvillea, and even grackles cawing.

From one lagoon to the next, through a small neighborhood park and on the nearby beach and weir, we were guided to count birds alongside seasoned Audubon veterans. I scribbled the names of birds being called out by the leaders and participants and fervently tried to keep up with the respective sightings. Ruddy ducks, red-shouldered hawks, white pelicans, brown pelicans, gulls, towhees, warblers, coots, and sparrows, just to name a few. I learned so very much, including the fact that despite all appearances of their jet-black feathers with white-tipped bills bobbing in the water, coots are, in fact, not ducks!

Each and every single bird was counted by our group and would later be tallied alongside the other groups who had participated in the CBC throughout the region that day. The count would then be included in the state's total, then to the national count, which would ultimately include much of the entire Western Hemisphere. The data collected is vital in ongoing conservation efforts, research, and scientific study and is a wonderful international tradition sponsored by Audubon chapters and affiliates to promote conservation and the accounting of as many birds as possible within a given time frame.

Just as vital is the gathering of persons interested in their communities and desirous to get involved with others to change the world for the better. Having a tradition to look forward to that invites and includes new souls is indicative of a sensitive stance and the willingness to extend beyond to others. Such good will and intention reverberated throughout my whole self that day. So much so that Tom took note of just how happy I seemed to be.

He didn't know it, but that moment of acknowledgment reinforced for me that I was where I was meant and needed to be, to feel and be with nature and among like-minded others. I imagined that this must be how one feels following a rebirth in baptismal waters, then going on to share the good news with others. In fact, at the end of the day, after all the counts had been tallied and the chili lunch enjoyed, I inquired about formally joining, and I continue to be a proud member of my local Audubon chapter.

When I am feeling low, I close my eyes and revisit that first outing to watch birds with others whom I didn't know. The other participants didn't know about my secret isolation at home, my certified disabilities that are mostly invisible to the eye, nor my desperation for healing. The Christmas Bird Count was the perfect decision, and I am grateful that I trusted the Universe.

It can be extremely difficult to go outside one's comfort zone and try something unknown and new. This is particularly true for those among us who are physically and mentally constrained due to chronic illness, unrelenting disease, and a lack of access to viable resources. Depression, stress, anxiety, shame, and fear, if left unharnessed, can imprison the mind and the body. When this happens, the opportunity for growth and healing is unintentionally forfeited. This then leads to feelings of isolation and lack of connection with family, friends, and, generally, the world—your intended world.

You can break this endless cycle by birdwatching beyond where you live, literally and figuratively, and exploring the healing power of engaging in your community. Each one of us is unique and has different needs and abilities, and we need to reach out and make connections, especially when we would rather not.

The pure contentment and positive healing effects from birdwatching through my windows, in the backyard, and eventually my neighborhood gave me the inspiration and the confidence to dare and consider more. From having an intense fear of birds to the joy and healing I have from watching them, being able to acknowledge this fear allowed me to overcome other obstacles in my life.

At this Audubon chapter event, no question was too basic, and talking and resting were not only allowed but encouraged. Tom's stewardship that day led to my eager return to join in countless more outings and learn more about birds and birding. Going to this Audubon event was, for me, literally showing up to help count the birds observed that day.

A Parliament, A Committee, A Kettle, A Chime, A Squadron

Many species of birds form flocks—on land, at sea, and in the air—but not all birds do. Flocks vary among species of birds, including their size, quantity, and purpose. Generally speaking, you can refer to a gathering of more than three birds as a flock, whether comprised of one species or different species and regardless of whether they are seen on a body of water, in trees and shrubs, on buildings and structures, or in flight.

Consider, for example, the enormous gatherings of European starlings in trees and in spectacular collective yin and yang of flight above rooftops, referred to as murmurations. Or consider flamingoes' perfected polonaises at water's edge, called flamboyances, which are displays of synchronized movement on a grand avian scale. Each bird relies on its neighbor and takes their cue as to their next step and so on, which creates a powerful sense of interconnection and of belonging to behold. From afar, their masses replicate a completely different living organism altogether and attest to the power of many over one. There are videos online that show murmurations forming a colossal bird in flight and others of massive waves ebbing and flowing as though an ocean in the sky created by hundreds or thousands of birds moving in unison.

Perhaps you've been amazed and delighted at first hearing and then eventually spotting the iconic V-shaped formation of boisterous geese flying in earnest overhead as though late for an important appointment, or the commanding glide of silent pelicans who seem to be on patrol along coastal shores and lakes. These are examples of birds who form flocks during migration, mostly in effort to efficiently conserve energy in order to travel great distances for longer periods.

Birds also form flocks to increase protection from adverse elements, provide more warmth, and ward off predation; potential threats cannot penetrate a massive fortress of unified birds nearly as easily as they can attack a single flyer. Flocks also enhance mating selection in congregates, called leks, where males strut their colorfully elaborate plumage in competition for female adoration. Think of the sage grouse found across the western sagebrush plains of the United States and parts of Canada in spring or the vibrantly colored manakins found in the tropical woodlands of South and Central America.

Some birds form communal flocks, which provide protection and added care for their chicks in rookeries, nesting grounds with nests situated closely near others to assist in navigation. Still others create flocks of symbiotic mixed species in order to optimally thrive. Mixed flocks of birds are highly efficient communities that capitalize on the differing capabilities and needs of the various species.[3] For example, while one species eats high up in the trees, a different species will forage for food on the ground; and while one species will alert others to predators lurking, other species intermingle, thereby increasing their numbers as a whole, which wards off danger.

Research supports that increased involvement in your community improves mental health and well-being.[4] This, coupled with the proven benefits of birdwatching[5] offers a wonderful foundation for ongoing involvement and support. It can be difficult to get started on a new journey, and being brave can often feel like misery, but we are not meant to be in isolation nor to go it alone. At least not for extended periods.

Participation at any level and within your capabilities and interests will restore vitality and heighten your sense of belonging. Once the first step is taken, your confidence will magnify your ability to continue on. For some, that may mean your first steps are to access social media and become acquainted with what birdwatching groups are available near you. Perhaps it's initiating a call to inquire about more information. For others it may mean asking for additional help from a family member or caregiver in order to expand your horizons away from where you live.

Being part of a community is at the very essence of belonging. Birds of a feather flock together, as they say, and birds of different feathers flock together as well. You can find strength in seeking a community that is reflective of who you are and what you value. You can also find strength from engagement in a community that is different from your own; such differences serve to celebrate your unique gifts and contributions in concert with every other person within the same community. When you recognize and acknowledge your neighbors' differences, you graciously welcome and accept them just as they are. In turn, such acceptance is reciprocated, one by one by one. This process of loving thy neighbor enriches your life and those around you. It grows and fosters a community that is safe and enriching for everyone.

Among us, in our families and with friends—where we live, work, learn, and play—we have many differences and a common desire for all to be well. Birds gather for the same reasons—to be bolstered against adversity, to provide optimal conditions for the ones they hold dear, and to maximize navigating through life.

The birding community is a microcosm of our broader world and societies, and is comprised of people with many differing strengths to contribute. They are actively growing their flock to embolden voices and reinforce a sense of belonging across racial and ethnic groups, gender and sexual orientations, persons with disabilities, among various ages and generations, and beyond. Inclusion and consideration for everyone is key to a healthy, well-functioning community and fosters a desire for more.

Just as birds are everywhere, so are we. I encourage you to consider your strengths and needs within this framework and override your natural tendency to remain in your comfort zone. Explore new options, reach out and make contact with another person or entity, and get involved at your own pace. The birding community encompasses a multitude of experiences. You can engage in volunteerism and share your expertise (we all have at least one), as well as reduce feelings of loneliness and isolation through routine participation in bird outings—trips nearby and to faraway lands.

Remember, during flock migration, the lead bird guiding the flock moves to the rear of the formation to rest and another replaces them to carry forward for all. That is to say, becoming involved with the birding community offers you the opportunity to share what you have and not feel as though you must know and do it all yourself. Increased frequency of active engagement enhances a heightened state of belonging. Being in nature and becoming connected to it, especially with others, leads to improvements in cognition, mood, cardiovascular health, vitality, and happiness.[6] When done consistently over time, it elevates your overall well-being.[7]

When we each do our part, we become part of a happier and healthier planet. You have options to find a good fit for you that is inspiring, fun, and free of judgment! Embrace taking the next steps beyond your nest.

GOING OUTSIDE IN YOUR COMMUNITY

Vital to a rewarding experience on a bird outing is a willingness to go beyond what is familiar in order to form connections rooted in the strength of differences. You are now ready to venture beyond your nest and surrounding environment. As you go about exploring possible birdwatching opportunities that may interest you, take the time to enjoy your process of discovery: the new information and skills you acquire, the people and wildlife you meet, and the newfound adventures you find connecting and gathering with others in your community.

Before You Go

Go online and search for birding, birdwatching, and nature-centered groups where you live. Research options in your immediate community first and broaden your search's scope over time. As you look up information, you will naturally stumble across additional information and resources that may interest you as well. Many organizations have local chapters, state entities, and national offices including National Audubon Society, Sierra Club, the Nature Conservancy, Royal Society for the Protection of Birds, Friends of the Earth, and Birdlife International. (See additional organizations in Resources.)

Make a list of organizations and communities that you are interested in finding out more about. Identify at least three and make notes of questions you would like answered.

Reach out and make contact via phone, e-mail, and/or social media. Extend beyond your comfort zone and ask the questions you jotted down to a representative of the selected organization. Contacting a minimum of three organizations is ideal so that you have the opportunity to make a decision that best meets your needs. Keep in mind that in many locations the organizations are reliant upon volunteers, so perseverance may be necessary.

Coordinate your visit to a desired location or site during a recommended time and event. It is very helpful to seek the guidance of the representatives with whom you communicate. They are best informed to assist you with having an enjoyable visit. That said, many places have guides and ancillary apps to facilitate going on your own as well.

Do virtual birdwatching of your community and region. Doing so will assist you in developing a connection and familiarity with the areas you are considering visiting. You can also watch videos that show the birds in your area so that you become familiarized with their markings. Once you are in your community on an outing, you will have more ease with identifying birds in different stages of development.

Preview trail maps, guides, and apps that are posted online. Take into consideration your physical and mental health needs, travel distance, and specific amenities that you require and/or prefer. (See Birdability in Resources.)

In Your Community

Refer to and incorporate the steps outlined in Chapter 1, Birdwatching at Home.

Review and/or compare trail maps, guides, and apps that are posted, if available. Supplement with other information you find online and through communicating with others.

Bring assistive devices and equipment.

Factor in your physical and mental health needs.

A willingness to make new connections on a regular basis is key.

STARTER TOOL KIT

An important tool you will need is your intentional effort to create new connections, deepen old ones, and engage in community.

Include items from the Starter Tool Kit in Chapter 1, Birdwatching at Home.

Preview your intended community organization and the surrounding environment.

Review trail maps, posts, and related apps.

Bring along assistive devices and equipment, as needed.

Prepare questions you may have for organization representatives.

Consider your needs regarding membership and opportunities to become involved as time permits.

EXERCISE PROMPTS

Enjoy your process of discovery as a precursor to making meaningful connections.

Explore online, print out, and attach a photo of a place you want to visit.

Research nature-based organizations near the place you have chosen. List them here:

___

___

___

Research nature-based organizations in your region:

- Local
- State
- National
- International

Now do this for other kinds of organizations that interest you.

What is one goal you have to participate in birding in your community?

What, if any, challenges do you have that interfere with your goal?

What steps can you take to reduce the impact of your challenges?

I met: ____________________________________

and we: ____________________________________

I belong to the following flock(s): *(include various associations, e.g., spiritual/religious affiliations, professional and/or regional communities, cultural/ethnic/gender identities and/or orientations, political identity)*

What do you value most from being a part of your flock(s)?

Birds I observed:

I am grateful for:

Reflection Pond

We cannot see our reflection in running water.
It is only in still water that we can see.

–Taoist Proverb

Answer the following prompts as though you are a crane standing at the water's edge, gazing into a still pond at your reflection.

Before I went on today's birding outing in my community, I was:

During my birding outing in my community,
I felt:

And now, I feel:

CHAPTER THREE

SOAR

Birdwatching as Mindfulness Meditation

A primal sort of scream pierces through the night's air, boldly interrupting desperately harvested restoration and hope-filled magical dreams. What had begun as an urgent care visit for dangerously high blood pressure quickly disintegrated into Fran being admitted for cardiac-related monitoring and eventual hospitalization for "significant cognitive impairment with behavioral disturbances." Then came an unexpected extensive hospitalization in a geriatric psychiatric unit until she finally returned home gravely ill with pneumonia and the flu, now unable to walk, sit up, or care for herself.

Night after night as she lay in her rented hospital bed in her apartment's small bedroom, my dear beloved mother's terror was often uttered in a man's guttural and full-bodied voice as though she were horrifically possessed. She had vivid hallucinations of starving little girls needing to be fed and large spiders crawling over the walls and floors. She was convinced of being served poisoned food and water, which rendered her suspicious of all and unwilling to eat, drink, and take medication.

Since the age of nine, Fran had stealthily protected herself from varying harms that attempted to visit themselves upon her, and she had learned to stand tall and resolute, which meant being prepared for battle both physically and psychologically. In her 84th year, that fierce little girl had returned.

Fran's lifelong devotion to the care and raising of me, my brother, my children, and some of their children has always been otherworldly in measure and an impossible standard to match. But as Alzheimer's and Parkinson's disease gradually ravaged her brilliant mind and her physical body with relentless precision and merciless cunning, I became the one who needed to show untiring devotion. It required a knowing of my true purpose and demanded that I rein in my errant ego so that I could be present with her in each precious moment, in the beautiful simplicity of her existence.

I slept on her sofa in her self-described "living room of many colors" adjacent to her bedroom. Several times each night I would be jarred awake by painful racing heartbeats that throbbed and pounded against my chest wall and temples, all in oddly synchronized unison, as if responding to Fran's disoriented screams. Each time I would guide my mother back to a rested sleep, deceptively assuring her all was well.

As days became months and then over a year of providing nearly round-the-clock care for my mother, my body's chronic ailments once again became exacerbated, leading to my own deterioration. At times, my mother had in-home support caregivers a few hours each week, which permitted me to briefly visit home. Usually, though, we could not afford private nursing, and so I wrestled with the challenges of my mother's care at the expense of savoring each moment we spent together. In the darkest of hours, my mind and body were beyond weary and in indescribable pain.

One early morning, following countless episodes of her night terrors, triggered bed alarms, and requiring assistance with bathroom trips, I went outside to sit on my mother's small deck and look for birds, just like I did at home, in my own nest. The longer I sat, the more birds I saw. Bejeweled hummingbirds, conspicuously masked orioles, and the ever-present house finches all made their appearances in the bottlebrush tree that grew right in front of the deck's railing.

I was struggling with my adequacy to provide the best care for my mother, day in and day out, and to honor her with the grace and endless patience she so rightly deserved. And I needed all that I had learned and experienced from birding at home and in the community to usher comfort and solace into my heart.

I asked the Universe for guidance, and immediately an Anna's hummingbird landed on the feeder and began to sip and sip and sip some more. It flew onto the railing's ledge and perched, facing me ever so still for what felt like an eternity, far longer than is customary for these mighty little birds. This tiny, feathered friend was a big, big gift, and I began to feel a warm gratitude fill spaces within me that had been hollowed out by sheer exhaustion, pain, despair, and fear. As I focused on the hummingbird, I stopped thinking about my circumstances as problems to be overcome. Instead, I was able to see them for what they were in the purest form, not good or bad, just what is now.

With practice, my ability to mindfully be with birds increased. During extremely difficult times when I was consumed by depressing thoughts, physical pain, and stress, I would step outside and search the sky. At times I would follow a hawk as it soared high overhead next to a vast canyon. As the bird rose on the air's currents, I would slowly breathe in and out in united reverence. The higher it ascended, the deeper my breathing became, allowing my body much-needed relaxation and alleviation of anxiety.

As the bird gracefully soared in concentric circles, I would close my eyes in order to actually *see* the bird in its purest form. Its iconic eagle-like calls confidently ricocheted from mountain slopes to rooftops as a validation of its existence, not good or bad, nothing more and nothing less. The longer I was attuned and saw the bird through my mind's eye, the more my other senses were intensified, including the recognition of my own existence, not good or bad, nothing more and nothing less. In those mindful present moments, thankfully, I became aware of what matters most in my life.

With my eyes now opened, both literally and figuratively, the distanced hawk soared on to a different part of our world. On days that permitted, Fran and I would sit outside together and gaze at the birds nearby in "her" tree and high above the nearby canyons. Other times, my mother connected with them inside from her recliner by looking outside at the bird feeder with contentment. Her feathered visitors would invariably make her smile with fond remembrances. Perhaps beneath the piled travesty of memory loss lay the buried treasure of revisiting life's moments, like a Buddhist's bell.

With and for my mother, I learned to soar.

Isn't it true that in the darkest depths of our despair comes illumination and answers to fretfully sought-after questions? The knowledge we gain manifests from and through our perceived pain, suffering, and lived experiences, which serve as rites of passage to deserved personal growth, inner peace, and ultimately, transformational healing. With this, we can soar above that which challenges our sensibilities to unforeseen and glorious heights.

Most often when we think of a soaring bird, the regal bald eagle or mighty condor comes to mind. Perhaps you have been fortunate to have witnessed such magnificence, these larger-than-life birds with outstretched wings effortlessly riding currents of air as they search for a meal. With pure awe, we watch the bird go extremely high and for such long distances and wonder, *how do they do it?* And then we dare to imagine, *what would it be like for me to be as free and soar so high above it all?*

Understanding how and why birds soar offers us an opportunity for guidance about our own quest to do the same—mentally, metaphysically, and spiritually. That is, mindfully.

A bird's ability to soar is predicated on specific physical characteristics and environmental conditions, including large and elongated wings with a specialized tendon that locks into place and an innate adeptness to harvest energy from air currents.[1] These air currents are sun-warmed and rise from the ground, are deflected skyward off cliffs and mountain slopes, and form high wind-speed gradients over oceans.[2] Birds soar for practical reasons: in hunt for food, to travel, and to avoid predation. They use most of their energy reserves when taking off, which requires costly wing flaps, but once they catch a thermal lift, they are then able to effortlessly climb higher and then higher. Hence the saying "Spread your wings and fly."

One of the largest flying birds on Earth, the endangered wandering albatross, is known to soar at sea for many years and never touch land!

Just as birds use their terrain to buoy themselves skyward, we can equally lift ourselves beyond our own seemingly unsurmountable valleys and mountains. Most of us weather obstacles and hardships using our resolve of inner strength and determination, support and encouragement from others, or a combination of both. These upswells of fortification carry us high above our challenges and bring about transformation. And through it all, we can learn to express loving acceptance of ourselves, of others, and ultimately find our own peace and healing. Regardless of how long it takes, we can soar.

From desperation, I learned how to soar with the birds and be free from pain, suffering, depression, and worry. Free to live in the moment and surrender to their respective realms of my thoughts about the vestiges of the past and prospects for my future. And the most precious gift of all, free to accept conditions and circumstances just as they are, along with acceptance of who I am.

MINDFULNESS MEDITATION

Mindfulness meditation is the practice of intentionally acknowledging our thoughts and feelings, and the environment we occupy. By doing so, we become aware of our embodied senses and what they present to us with loving acceptance and understanding. This cultivates compassion for ourselves and for other living beings.

Now that you have spent time watching, listening to, and being with birds, I invite you to deepen your connection with them. Your feathered friends can serve as your guide to staying in the present from moment to moment. Allow the natural paces of nature to help guide your time and place.

Allow the birds to help you achieve and maintain improved mood, increase your ability to focus and decrease your pain, and develop genuine compassion for yourself and therefore for others. You can develop fulfilling inner peace by incorporating the principles of mindfulness and meditation while being with birds. I welcome you to join me and learn how to symbolically soar high above your darkened clouds of burden and worry, and revel in your healing mind's communion with nature. You will come to know the birds and yourself through a different way of healing right alongside them.

GOING OUTSIDE AT HOME

Although mindful meditation can be practiced anywhere and at any time, whenever possible, it is most ideal for the practice of mindfulness meditation with birds to be done among them in their natural outdoor environment, including yards, parks, forests, lots, beaches, and trails.

Throughout your mindfulness meditation session, I strongly advise you not to use equipment to see or hear the birds unless it is an assistive device. Mute your phone and leave your binoculars, scopes, field guides, and other birding apparatus behind. Only rely on your natural capacities.

I have carefully outlined the steps to practice mindfulness with birds for ease in getting started. A typical session lasts approximately 15 to 30 minutes, though it is completely at your discretion. As you become accustomed to it, you may find that the actual practice will take less time.

Let's begin.

Step One

Walk slowly and respectfully into your outdoor space as though you are an unexpected guest.

Stand still on the perimeter of the area, preferably with your back against your home, building, or structure. Your "yard" includes any and all neighboring space within your extended visual field.

Stand for at least one full minute, breathing as you normally do.

Use your senses to become acclimated to the environment:

HEAR first, listening for **the sounds while keeping your eyes open.** This step is important, as it actively engages your mind to assess auditory stimuli: the rustle of tree leaves in the wind, chimes hung to reveal the air's presence, melodic birdsong as it floats by, a plane scraping high up in the clouds, children's chatter, an appliance's whir, the battering of construction, bus squeals. Try to pinpoint where each sound is coming from using your ears like a dowsing rod.

LOOK around generously, in all directions. Sweep up your eyes, beginning from the ground toward the skyscape. Then look from right to left, left to right, as far as your vision can extend in all directions.

Then look diagonally across the space. Begin at the upper left quadrant (sky), looking diagonally down to the lower right quadrant (ground). Repeat beginning from the upper right quadrant (sky), looking diagonally down to the lower left (ground).

What do you see? Dirt, trees, clouds, houses, rooftops, birds, people, cars, light posts?

While you may see birds, and hopefully that is in fact the case, this initial act of looking is intended for you to take in all that your environment has to offer without any limitations, parameters, and expectations of the process.

Turn completely around to face your home and repeat.

Repeat this again while standing facing right, and again while facing left.

Take a mental note of all you have seen in as much detail as you can, yet avoid lingering too much. Just take note of what you observe and notice what else you hear, see, or feel.

FEEL the air that surrounds you. Do you feel the breeze or wind against your skin? Do you have sensations of warmth, cool, freezing cold, or heat? Do you feel moisture, wetness, or dampness, stings from penetrating sunlight or bitter cold?

Our sense of touch is often taken for granted, but it is vital to becoming connected to that which is within reach.

SMELL by breathing in and out through your nose as you normally do.

Take note of what you smell around you: grass, plants, pine, car exhaust, barbecue smoke, laundry, the pure air itself.

TASTE what is in your mouth.

We often dismiss the power of taste to inform us about our environment beyond food and beverages. I encourage you to explore what is around you through taste, much like an unfiltered and curious child would do—metal, concrete, stucco, leaf, bark, bamboo, plastic.

(*Optional*)

Place the tip of your tongue ever so briefly on a handful of things in your surroundings. Move your tongue gently to the roof of your mouth and notice the flavors or possible absence of flavors. What do you taste? Salty, bitter, green, sweet, metallic, bland, spicy, earthy, woody, plastic?

I encourage you to taste and/or consume items only if you are knowledgeable of their content and effects.

Find your "YOU ARE HERE" within the space that you are standing. Are you at the center, on the edge, or oriented at a corner position? Take a mental note of where you are standing in relation to everything around you. Identifying your spatial orientation, or your place in the world, will help ground you by putting all that exists in the context of your current perspective.

Step Two

Now decide where you want to sit or lie down. Find a spot that is comfortable and that lends itself to you spending time on it. Select your spot with intention and be free of judgment about your choice. Allow your inner compass to guide you to what it is you need to experience each time.

SIT or lie down any way that you choose. There are no better ways than others. Allow your body to determine how and where to be positioned that seems right for you.

For example, you may select a chair, stool, or bench, or you may prefer a flattened boulder, ledge, wall, grassy area, or being on the ground.

Determine how you wish to sit, such as legs crossed or lotus style.

Adjust yourself to achieve maximum comfort, which may involve:

Both feet flat upon the ground

Being postured with your back upright, though not stiffened

Arms gently to your side or in your lap, unrestricted and relaxed

Palms of hands positioned in an open, inviting manner, allowing them to fall naturally as they may

Simply SIT and simply BE.

You now have a new vantage point from which to observe. Settle into being in your space and actively take in all that you are experiencing.

I suggest that you sit and be for a little while longer than you may initially plan, particularly if you are practicing this for the first few times. It can take time for our bodies and minds to adjust completely to the practice of being, free of a prescribed agenda.

Use the same steps you did when you first stood outside:

With your ears. . . . . **HEAR**
With your eyes . . . . . . **LOOK**
With your skin . . . . **FEEL**
With your nose . . . . . . **SMELL**
With your mouth. . . . . **TASTE**

Begin to specifically look for any and all birds. Part of becoming mindful is granting yourself permission to indulge your curiosity. Scan your chosen space for birds. Ponder why the birds are doing what it is they are doing or behaving in the manner that you observe.

Give yourself at least one minute once you find birds.

Gently close your eyes, sit, and be.

Continue sitting with your eyes closed and notice the natural rhythm of your breathing as you acknowledge what you:

HEAR, LOOK, FEEL, SMELL, and TASTE.

Deepen your attention to your breath.

Think and repeat to yourself: *IN* breath, *OUT* breath, in unison to your natural breathing pattern.

Mindful breathing quiets our internal chatter, or "the noise in our head." The more you focus on your present act of breathing in and breathing out, the more a sense of calm will be cultivated, and your body will relax. Continue your mindful breathing until you feel completely calm and receptive to the next step.

Acknowledge the birds you hear all around you, far, far in the distance, midrange away on buildings and lawns, and in gardens, and finally, nearest you.

With curiosity, ask:

- How many birds are there far away, at a mid-distance, and near?
- What are they doing and why?
- What else piques your curiosity?

Choose one bird to focus on and deepen your mindfulness meditation. Try and do this with as little self-analysis or discernment about the process as possible, and rely on your natural internal compass to guide your decision.

Accept the birds as they are, intended for you here and now, winged gifts to offer you what you need at this precise moment. Nothing more, nothing less. Allow the bird to select you too. You will just know.

Sit and be with your chosen bird, keeping your eyes gently closed.

Give inner voice to breathing *in*, breathing *out*, as you hear the bird.

In time, you will experience a synchronistic convergence between your mindful breath and the bird's song. The natural world shares rhythms and patterns that permit magical connections. It is that which we aim to join in on to restore ourselves to a more elemental level.

Step Three

Use the same steps as before, except this time you are now able to further deepen your mindfulness meditation by keeping your eyes closed.

With your eyes *closed* . . **SEE**
With your ears . . . . . .**LISTEN**
With your nose . . . . . **INHALE**
With your skin. . . . . **ABSORB**
With your mouth. . . . .**SAVOR**

See

See the bird through your mind's eye.

Perceive the bird and envision its form in your mind. Focus your thoughts about what the bird looks like and what it is doing in the center of your forehead, between your eyes, as though you are looking from within your mind, facing outward.

The way in which you see the bird in this way strengthens your spiritual connection to it, the space you are currently in, and to yourself. Your perception is unique just to you and offers sacred connection from within.

What do you see?

Listen

As your attention drifts off to other sounds, and it inevitably will, simply acknowledge this process and gently direct yourself back to your bird and the layers of sound it makes using your focused attention.

Inhale

Unlike the act of naturally smelling what is in your environment, I invite you to breathe in slower and more fully from the bottom of your lungs upward—a yogic breath, if you are able. If not, then try to inhale slower than you would normally breathe, but not in a way that causes you discomfort.

As you inhale, acknowledge all that you can and envision the scents floating along on your inhaled breath and entering your lungs and diffusing throughout your bloodstream into different body parts and organs.

Absorb

Give purposeful notice to the sensations your mind and body experience, right now at the precise moment as you share space with your bird.

Is what you absorb shared by your bird? What are the sensations you are absorbing?

Savor

Savor the flavors of your mouth as it is and try to differentiate them, just as you did when choosing your bird to spend time with.

What are you savoring?

Flavors can be unabashedly straightforward and, at other instances, layered with complexities.

Are they simple or complex?

Your internally channeled senses to see, listen, inhale, absorb, and savor are guided by the bird's movements, sounds, and its own senses. Once it departs, sit and be with your experience. Some gifts are brief and fleeting, while others remain and accompany you for a while. However long it may be, it is what you need for now.

The process of accepting how birds present themselves to you without preconceived expectations is your gateway to self-acceptance. As you continue to practice mindfully accessing your senses with birds to harness this harmonious balance, you will nurture understanding about them and you. By going beyond marveling at the birds' plumage and practicing mindful meditation with them, you commune with the birds in compassion and loving-kindness, which will in turn give you the capacity to experience the same with yourself and others.

By creating this sacred space to be present in your life, you claim mastery of your mind and body, and now, dear friend, you soar.

STARTER TOOL KIT

You can practice mindfulness meditation at any time and in any place of your choosing.

The most important tool in your kit is *the gift of time.* **Set aside at least 15 minutes**, ideally 30–45 minutes, for each session. *Leave all birding equipment behind and phones muted.*

Choose a space to **sit and be**

Something to sit on (*optional*)

A journal and pen/pencil to write and/or sketch your session afterward (*optional*)

Weather-specific gear (jacket, umbrella, hat, sunglasses, sunscreen, specific footwear)

A day pack for mindfulness meditation sessions away from home

FROM INSIDE LOOKING OUT

Not being able to go outside doesn't preclude you from being able to practice mindfulness meditation with birds.

Choose a window or doorway that can be opened and offers you the best vantage point to view your outdoor surroundings.

Stand, sit, or lie down as still as possible near the window or doorway. Birds are very aware of movements, including from inside where you are.

Stand for at least one minute, breathing as you normally do.

Use your senses to become acclimated to the environment:

HEAR first *with your eyes open.* This step is important as it actively engages your mind to assess auditory stimuli. Try to pinpoint where each sound is coming from.

LOOK around generously, sweeping your eyes from left to right, up then down. Sweep up, beginning from the ground and moving upward toward the skyscape; look from right to left, left to right, as far as your vision can extend in all directions.

Then look diagonally across the space. Begin at the upper left quadrant (sky), and look diagonally down to the lower right quadrant (ground). Repeat beginning from the upper right quadrant (sky) and look diagonally down to the lower left (ground).

What do you see? Dirt, trees, clouds, houses, rooftops, birds, people, cars, light posts?

While you may see birds, and hopefully that is in fact the case, this initial act of looking is intended for you to take in all that your environment has to offer to you without any limitations, parameters, and expectations of the process.

Take a mental note of all you have seen in as much detail as you can, yet avoid too much lingering and time.

FEEL the air that surrounds you. Our sense of touch is often taken for granted, and yet it is vital to becoming connected to what is within reach.

Do you feel the breeze or wind against your skin? Do you have sensations of warmth, cool, freezing cold, or heat? Do you feel moisture, wetness, or dampness, stings from penetrating sunlight or bitter cold?

SMELL by breathing in and out through your nose as you normally do.

Take note of what you smell around you: grass, plants, pine, car exhaust, barbecue smoke, laundry, the pure air itself.

TASTE what is in your mouth.

We often dismiss the power of taste to inform us about our environment beyond consuming food and beverages. I encourage you to explore what is around you, much like an unfiltered and curious child would do—metal, concrete, stucco, leaf, bark, bamboo, plastic.

(*Optional*)

Then, with the tip of your tongue, place it ever so briefly on a handful of things in your surroundings. What do you taste? Salty, bitter, green, sweet, bland, spicy, earthy, woody, plastic?

I do not encourage consumption of any part of these items unless you are knowledgeable of their content and effects.

Find your "YOU ARE HERE" within the space that you are standing. Are you at the center, on the edge, or oriented at a corner position? Take a mental note of where you are standing in relation to everything around you. Identifying your spatial orientation, that is, your place in the world, will help ground you by putting all that exists in the context of your current perspective.

Position yourself to be able to sit comfortably in the manner that you choose. For some that may be on a chair, stool, or other type of furnishing, and others may prefer the floor or doorstep.

A GUIDED MINDFULNESS MEDITATION WITH BIRDS

This guided meditation can be done in any setting with your eyes open or closed.

Take a deep, slow breath in, hold for four seconds, breath out, exhale slowly to a count of five. Repeat with each passage.

Say aloud to yourself (and consider placing one or both hands on your heart):

> As I sit with the bird, I am free. I listen to the bird; I listen to myself. With my eyes closed, I see the bird; with my eyes closed, I see myself. I inhale deeply the scents that envelop the bird, into me. All that there is surrounding the bird, I absorb in me. The bird departs; I remain, in peace.

EXERCISE PROMPTS

I am:

inside outside

The space I chose to meditate in is:

Describe the bird you spent time with *(include features and behaviors)*:

What did you look at with your eyes open? Then see with your eyes closed? What do you now see within yourself?

What gifts did you receive during your time practicing mindfulness meditation?

If I could be any bird, I would be a:

Why?

When I ______________, I soar.

Reflection Pond

We cannot see our reflection in running water.
It is only in still water that we can see.

—Taoist Proverb

Answer the questions as though you are a crane standing at the water's edge, gazing into a still pond at your own reflection.

Before I began my mindfulness meditation session, I was:

While I was practicing my mindfulness meditation, I experienced:

And now, I am:

CHAPTER FOUR

BIRD LEGS

Birdwatching and Your Health

I could hardly even hold a glass of water, brush my teeth, nor pick up a stray dime without my hands shaking uncontrollably. They served as my constant reminder that I was not okay. Pain and physical challenges can be quite insidious and creep into the tiny cracks and crevices of even the sturdiest among us. The inability to control one's body and functions often breeds fear, despair, and shame. Following complications from an unexpected hysterectomy, my body screamed *"No!"* and shut down.

Consultations and follow-ups led to more consultations with countless specialists. I was desperate for help and dutifully took all treatments and medications against

my personal objections. Days turned into weeks, which turned into months of what felt like my new full-time job going to doctor visits and undergoing treatments that left me stuck at home, "couch-bound," and in unrelenting pain. It became my way of life. The strong doses of prescribed medications left me to navigate a thick haze of intense fatigue, extended daytime sleep, and an overall sense of unwellness. My memory, which had always been sharp, had dimmed to a low ember. Pajamas became the day's clothes, as taking a shower and changing into work clothes were now a thing of the past. Every day I spiraled deeper into the welcoming grip of depression.

As the months became years, my condition was ultimately diagnosed as chronic regional pain syndrome (CRPS). I was told by an expert that there was no cure and all he could offer me was participation in an experimental research study in Los Angeles, which involved the surgical implantation of a medication pump into my chest with no guarantees. I went home and into bed for days, often mustering just enough strength to help my children ready themselves for school.

The realization that I truly would continue to worsen over time was devastating, and one of the hardest parts to reconcile was my job. I had prolonged giving my employer final notice, but it had become inevitable. I had been with the nonprofit agency for over 20 years, since I was 23 years old. I'd started on the night shift providing supervision and care for at-risk teenage girls and worked my way up to program manager over the years. I had grown up in the positions at work and developed several close friendships. The countless hours and responsibilities of managing clinical treatment and services for some of the most vulnerable souls was who I knew myself to be: responsible, goal-oriented, a recognized achiever against perceived obstacles, and one who never gave up. Yet giving up was all I could think about now.

I became starkly aware of the changes occurring around me. Few friends called anymore and even fewer visited, which prompted an existential reckoning about the human condition and my actual relevance. I understood that life was going on without me. I felt hopeless and sorry for even entertaining the possibility that my health issues could ever improve. Hope for a better way had all but vanished, but the fighter deep down in me didn't want to give up. Then unexpectedly, healing flew in on the wings of feathered angels.

Though my legs were literally weakened, unconditioned, and often failed to cooperate, I became strong in spirit from gradual immersion outdoors in and among nature. Little by little, I began to feel physically restored and strengthened. I could sit outside for longer periods, measured in increments of one and two minutes. I began walking just a bit farther on some days. Other times, I became so overly exuberant and ambitious from the joy of watching and being with the birds that I would overexert beyond what was reasonable for my current functioning. This would then lead to undue additional pain, strain, and more rest required to get back to my baseline.

The longer the challenges of maintaining decent health—let alone good health—continue to go on, the further into a kind of magnetizing darkness many unwittingly descend. We, those of us with medical and physical problems, are knowingly our own harshest critic. We strive to be "normal" and okay, to fit in and step up in defiance of what our mind and body truly require and need. And that's not to say that we ask for nor want well-intended yet belittling pity. No. We are asking to be listened to, understood, and believed, and nevertheless, respectfully included.

This includes those among us whose outward appearances make us seem to be in fairly good health and form. Yet eventually one or more cracks widen to reveal our flaws. The shock of such flaws can create alienating disbelief and skepticism and compound our already mighty struggle.

This is mirrored in the world of birds as well, and is usually questioned from the sight of a bird's legs. Bird legs, much like our human legs, come in varying shapes, colors, and sizes. A bird's leg is characteristically unique among mammals, as it is structurally very strong given their synsacrum, which is their fused pelvis connected with the spinal bone.[1] Birds have three distinct segments that make up their legs, and most walk on their toes. The portion of the bird's leg we are accustomed to viewing with its characteristic backward bend is actually the bird's ankle. Its knee and upper femur are most often hidden under feathers.

Most birds have four toes: three that are forward-facing and one facing backward in hind adapted to facilitate perching. Exceptions to this are the ostriches, rheas, and emus that stand so very tall on strong, thick legs and only two toes. Conversely, the very small swift, which acrobatically darts and tumbles in the air, has no need for legs and uses its tiny toes to cling to its habitat along cliffs' edges. The distinct leg differences among bird species are indicators of the habitats that they navigate and their capabilities.

Despite their differences, the legs of birds all serve to provide strength and stamina. Their legs buoy them above water as they dabble among eelgrasses for food, propel them to run and forage, clutch prey, shelter eggs, moderate their bodies' temperatures, and preen their feathers clean. With rare exception, bird legs are hollow to afford effortless flight and ease of locomotion through the water.

WELL-BEING

The connection between nature and well-being is fundamentally organic. As a species, we are naturally inclined to want to be in environments infused with fresh air, plants, trees, soil, water, and animals. And yet we often find ourselves disconnected, paying homage to modernization and its technology with incessant demands that anchor us inside artificial surroundings most hours of each day. Those of us who experience the compounded challenges of less-than-optimal health or have disabilities may find that we regrettably spend even more time indoors for various reasons: physical debilitation, isolation, habit, and the comforting familiarity of our indoor surroundings.

Well-being is associated with measures of physical illness, mental-health functioning, access to resources including green and blue spaces (for example, trees, gardens, and water), longevity, social connectedness and environment, and one's own self-perceived determination of their health.[2] Oftentimes without realizing it, we take stock of how we feel and adapt our behavior in accordance with our personal self-assessments. Pain can interfere with rational processing of our thoughts and emotions, and such distortions contribute to additional suffering. Whatever it is that we are struggling with physically deserves to be acknowledged for what it is in its purest form without our tendency to tamp down our symptoms to make it less burdensome for others or to attempt to fool ourselves. Doing so minimizes who and what we presently are.

You gain your symbolic legs of strength from standing in your truth of self-acceptance of your current circumstances, good and bad, crooked and straight. From this empowered stance of self-acceptance, you have the opportunity and a reference to become aware of and experience profound growth and deserved genuine healing. In other words, in order to know where you are going, you need to acknowledge the conditions and circumstances you find yourself presently in. Nature is free of judgment and discernments. You can spend time just the way you are, and it is good enough. Nature is accommodating, adaptable, and encouraging. Take guidance from the wise and true—take guidance from nature to steady your journey.

Spending Time in Nature Is Healing

The benefits of spending time in nature include decreased levels of stress, high blood pressure, heart disease, cancer, and diabetes, and increased levels of energy, creativity, immune system functioning, and connectedness. Recent studies cite similar beneficial effects from time spent in nature with birds.[3]

In a study of 20,000 people led by researcher Mathew White of the European Centre for Environment and Human Health at the University of Exeter, it was determined that those who spent at least two hours a week, whether all at once or incrementally, in parks and other outdoor settings reported a significant increased sense of good health and psychological well-being.[4]

Birds are all around us a majority of the time in one way or another. Ideally, you should venture outdoors in their natural habitats and spend time watching and listening to them. For those of you who are unable to do so, you can still gain beneficial healing from inside where you live, looking out through a window or doorway.

CHRONIC PAIN

It's important for you to understand that chronic pain is very different from regular pain that is associated with a typical injury. Chronic pain is ever-present and lasts longer than six months. It assaults your nervous system repeatedly with error messages of injury as though you are repeatedly sustaining it for the first time despite the originating cause having been remediated through healing or simply going away on its own.

Chronic pain has a myriad of causes, including infections, back conditions, arthritis, cancer, adverse post-surgical reactions, and various pain syndromes such as chronic regional pain syndrome (CRPS), chronic fatigue syndrome, fibromyalgia, and inflammatory bowel disease (IBS), among many other causes. At the core of the endless sensations from pain is the nervous system alerting your brain and the rest of your body that an injury has just occurred that requires immediate attention.

In 2018, the Centers for Disease Control and Prevention (CDC) reported a revealing 20.4 percent (50 million) of adults in the United States has chronic pain, and 8 percent (19.6 million) suffered from "high-impact chronic pain." Also, they found worldwide that 1.5 billion people have chronic pain that directly correlates with increased symptoms of depression and anxiety, limited mobility, dependence on opioids with only 58 percent efficacy, and a general sense of poor quality of life.[5]

Those of us plagued with chronic pain know all too well how difficult it is to really describe what it is we physically experience each day. Pain is inherently understood, and enduring such pervasive and persistent pain causes emotional imbalance in the form of depression, fatigue, irritability, anxiety, fear, and disability for many. Pain can interfere with rational processing of thoughts and emotions, and such distortions contribute to additional suffering.

You can support your own healing from this cyclical pain process in part by first granting yourself permission to be *not well*! It is essential that you acknowledge to yourself that you have pain and feel unwell, and perhaps even feel alone, in the mire of debilitation. This recognition alone can begin your process toward improved well-being.

Spending some time each day watching birds will help you to experience a reduction in your pain levels, because the act of birdwatching in and of itself serves as a beautiful and welcome distraction. Slow, gentle movements performed within your capabilities help to ease chronic pain. Spending time feeding the birds, being outdoors, and near open windows and doorways are all ways you can help ease your health challenges. There are an increasing number of trails and park paths that are accessible for those who use wheels and assistive devices. (See birdability.org.)

Chronic pain and conditions can feel worse in isolation. Consider participating in bird outings and volunteerism through nature groups and with friends and family for the fellowship, encouragement, and, best of all, the pure enjoyment of it. For those unable to venture out, consider supporting nature-, environmental-, and conservation-based organizations by providing your skills and knowledge from where you live at home.

STRESS

The National Institutes of Health's National Center for Complementary and Integrative Health (NCCIH), describes stress as "a physical and emotional reaction that people experience as they encounter changes in life."[6] The American Psychological Association similarly defines stress as "the physiological or psychological response to internal and external stressors"; it affects every system in the body and influences how we feel and therefore behave.[7]

Despite being an accurate explanation, reading the definition of stress may seem to deconstruct it to the barest of meanings in light of the magnitude of its impact in our daily lives. Scientists have continued to assert the health benefits of some stress: to trigger the fight-or-flight response, flooding our bodies with the stress hormone cortisol in times of danger and emergency, for immune system regulation, and to reduce inflammation.[8] Some of us encounter stress and related stressors more so than others as a result of historical and generational inequities, environments, and related socioeconomic factors. Experiencing stress and stressors is inevitable, and yet we are consumed with trying to avoid it at all costs. This is due in part to societal norms that suggest a good life is a stress-free one and because many of us have not had an opportunity to learn effective strategies to cope with it. Learning what stress is and its symptoms is a huge step toward healing.

It is when stress is longstanding and overwhelming, negatively impacting our ability to properly function and cope, that it's called chronic stress, which is so very debilitating. We often endure it when under financial strain, relational difficulties, work and social demands, and hostile environments with violence, toxins, and limited access to resources.

Chronic stress can affect our overall well-being and all systems of our body, including musculoskeletal, respiratory, cardiovascular, endocrine, gastrointestinal, nervous, and reproductive. Even in the absence of a stressor, remembrance of it can induce heightened stress responses, both physically and psychologically, time and time again. Physiological symptoms of chronic stress include insomnia, head and stomach aches, high blood pressure, heart attacks and strokes, bowel changes, muscle pain and body aches, sexual dysfunction, and increased risk to developing viruses and infections. Frequently experienced psychological symptoms from persistent exposure to stress and stressors include depression, anxiety, irritability, and other mental health conditions.[9]

Having an awareness of what and how stress impacts you is key to developing effective strategies to combat the effects. And growing scientific evidence indicates that immersion in and with nature can truly serve as a reliable antidote.[10]

Exposure to green and blue spaces where there are trees, shrubs, and gardens within 10 minutes of travel time has been shown to help reduce stress and its related affects, including levels of cortisol, blood pressure, and glucose. A promising study conducted by a collaborative team in the United Kingdom from the University of Exeter, the British Trust for Ornithology, and the University of Queensland, suggests that even the act of observing birds and nature can reduce levels of stress. They learned that it did not matter what species of birds one watched. Rather, the quantity and the time of day were most vital to reported increased well-being and reduced levels of stress. The more birds observed, as well as doing so in the afternoon, yielded the highest correlation with significantly reduced stress levels, feeling connected to nature, and feeling happier.[11]

MaryCarol Hunter and her colleagues similarly conducted an eight-week study of participants' time spent in outdoor spaces, referred to as Nature Experiences (NE) or "nature pills," for a minimum of 10 minutes each time. Participants in the study were told to spend time in nature, then an analysis of the participants' saliva was conducted. It revealed that spending time in nature resulted in a reduction in the stress hormone cortisol.[12] A 20-to-30-minute "nature pill" showed the most efficacy and serves as further guidance for us.

Spending time outdoors in nature for up to two hours, even in the afternoon, watching birds, regardless of whether you can identify them or not, is restorative and healing. While the benefits of spending time in nature have been known throughout time and civilizations, concerted efforts to quantify them is growing given the drastic increase in health-related complications all over the world. Physicians worldwide are rallying for nature to be an integrated component of health and healthcare systems like Park Rx America in the United States, Nature Prescriptions in Scotland, which is endorsed by the Royal Society for the Protection of Birds (RSPB), and Healthy By Nature in Canada. They prescribe written prescriptions for patients to spend time outdoors and track the reduction of chronic illness, risk factors, and the significant health improvements in their patients' lives.

This good news extends to those who may have less mobility or cannot be outdoors. A recent study determined that benefits to your health can be obtained from watching nature and wildlife, including birds, from inside and on-screen. Whatever you are able to do and however you are able to do it, it is good enough!

Connecting to the birds near you is close at hand in some form or another. Special consideration to your health conditions and needs can optimize your time and make your experience more enjoyable. Listen and watch for your feathered friends to continue your healing journey.

Before You Go

Preview your destination via phone call, websites, and other online resources whenever possible to ascertain:

Facilities available on-site. Not all locations offer the same amenities. It is a good idea to have an idea of what you can expect in terms of food and meals, meetings and programming space, and educational and interactive offerings.

ADA-compatible and accessible accommodations for you and/or a companion. It cannot be assumed that they do or do not have necessary accommodations. Determining what will be a good fit prior to going will enhance your enjoyment and time spent there. (See birdability.org for more info.)

Bathroom availability and type based on your personal and companions' needs. For example, some locations offer larger bathrooms for families.

Parking availability and location to your destination.

Terrain of trails and paths. Some locations may be deceiving in terms of level of difficulty and accessibility.

Rest stops and benches/seating options. Consider your needs based on the estimated time to travel a trail or outing. Consider adding additional time based on your capabilities. Knowing where you can safely rest can be helpful.

Cell phone service. Take along your phone in case you need to call for help. If possible, check in at the ranger station, bulletin board, or other means to let others know your bird outing plans, particularly when out alone.

Weather conditions. Consider your health needs and capabilities in different elements and dress accordingly.

At Your Birding Location

Review and/or compare trail maps and guides posted, if available. This will assist you in determining terrain, duration, rest stops, and distance.

Assistive devices. Consider your personal needs and additional equipment that can increase your enjoyment. You may require more equipment than you are unaccustomed to in order to experience more of the environment.

Pack smart. Consider taking only what you need in order to keep your load as light as possible.

Water and snacks. Bring along food and adequate quantities of water for yourself and companions. It may become necessary for you to have additional nourishment to maintain your stamina.

Appropriate supportive footwear, hat, and clothing. Consider the duration of your trip and your physical health needs.

Weather conditions. Plan for changes in conditions and your personal needs.

Medication. Factor in the length of time you plan to travel to and from your outing as well as the time spent during the activity.

(Optional)
Chair or cushion. Easily portable, sturdy, and supportive.
Blanket
Birding companion

STARTER TOOL KIT

Include items from the Starter Tool Kits in Chapter 1, Birdwatching at Home and Chapter 2, Birdwatching in Your Community.

Preview your bird outing online and on-site.
Pack smart and include assistive devices.
Bring medications, water, and snacks.
Bring emergency contact information.

(Optional)
Chair or other supportive device for rest
Blanket
Whistle

EXERCISE PROMPTS

Date/Time:

The bird outing I plan to explore is:

I plan to spend ____ minutes/____ hour(s).

My health is important. I plan to support my venture outdoors by bringing along:

Although I am challenged by ________________, I have the physical ability to:

Bird legs are:

My health is like a bird's legs. I am:

My current physical health is:

Excellent Very Good Good Fair Poor Very Poor

Physically, I struggle with:

Physically, I used to be:

I accept that now, physically, I am:

PHYSICAL HEALTH SELF-ASSESSMENT

There is strength in recognition,
acknowledgment, and acceptance.

You can use this Physical Health Self-Assessment on the next page in two different ways:

1. As a daily journal: Fill it out on a regular basis.

2. Fill it out twice: once *before* and once *soon after* you have spent time watching birds.

Physical Health Self-Assessment

Today's Date/Time:											
My current physical health is:											
Well-being:											
Circle one:	0	1	2	3	4	5	6	7	8	9	10
		no symptoms			some symptoms				many symptoms		
My symptoms/feelings are:											
Chronic Pain:											
Circle one:	0	1	2	3	4	5	6	7	8	9	10
		no symptoms			some symptoms				many symptoms		
My symptoms/feelings are:											
Stress:											
Circle one:	0	1	2	3	4	5	6	7	8	9	10
		no symptoms			some symptoms				many symptoms		
My symptoms/feelings are:											
Other condition:											
Circle one:	0	1	2	3	4	5	6	7	8	9	10
		no symptoms			some symptoms				many symptoms		
My symptoms/feelings are:											

Reflection Pond

We cannot see our reflection in running water.
It is only in still water that we can see.

–Taoist Proverb

Answer the questions as though you are a crane standing at the water's edge, gazing into a still pond at your own reflection.

Before I went on my bird outing today, I felt (*physically*):

While on my bird outing, I felt (*physically*):

After my bird outing, I feel (*physically*):

CHAPTER FIVE

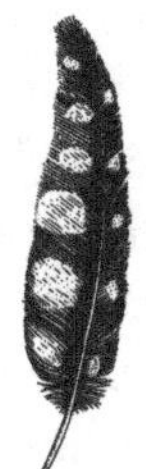

FEATHERS

Birdwatching and Relationships

Snow still blanketed the iconic Colorado mountain ranges and poured into the valleys like sugar spilled from a bowl. Its sunlit crystals sparkled amid the regal pines as we drove east along the interstate. Ski lifts and ski slope paths marked with ribbon and brightly colored posts dotted through forest clearings, ending where they began, at the lodges below. And then we both saw it, my husband and I, seeming to appear from nowhere: a very dark-brown bird with an impressive wingspan soaring down from the skies facing us heading west.

It descended low into the valley, stretching alongside the interstate, and glided just above the river that rushed over glistening boulders and lost tree limbs. As it glided

westward above the water, we drove eastward, slightly elevated, and eventually we could see its distinctive white head, yellow beak, and magnificent brown wings stretch over nearly the entire width of the water. We had never witnessed such a glorious sight in person before.

Almost instantaneously, we knew this was a gift from the Universe on behalf of my dad, guiding us with lifted spirit onward to our journey in his commemorative honor. His favorite birds were eagles, most especially the majestic American bald eagle, which he revered for its symbolism of freedom for the country. He valued it so much that he named his mountain cabin up in the Rockies the Eagle's Nest, because eagles could be seen up there, and it sat atop a valley that was free and full of big country sky, untarnished by city and lights.

My dad, a steady anchor in my life, had defied his sense of permanence as a founding member of our family—always there, conservatively predictable, and steady—and we were on our way this last time to prepare for his military memorial service in the coming days. *Thank you, Dad.*

We were silent in reflection, and memories beckoned. The summer before changed the depth of the significance of the Fourth of July for me and other family members forever. I now visit my eldest grandson's grave under the shade of a sturdy pepper tree on this national holiday filled with celebratory fireworks and get-togethers commemorating freedom and independence, which is so befitting given his beloved zest for life and adventure, as his accident was on this tragically prophetic day. Nearby, aged pine and oak trees serve as emblematic apostles for the residents and for those, like me, who come for a while. In his 16 years, his spirit was so strong for so many. Some visits are easier for me than others, but all are a gift. Sometimes I need to know that he is still near and not somehow lost up there without us, his family and friends.

On the occasions that I request his presence, I ask him to commune by way of the hawk, one of his favorite birds. Each time that I do, within a few minutes, a red-tailed hawk appears and perches high atop the same tallest oak near where I stand. I am certain it is him because the oak tree is the symbol his great-grandmother Fran bestowed upon him when he was young. Other grand- and great-grandchildren are trees and flowers too, but it was apparent that the oak fit him and his strong and ever-reaching personality. I only request his presence sometimes because I don't want to bug him, as I know he is a gregarious and generous soul who was always going somewhere and tending to the needs of others when here. I imagine him busy doing so now. I am comforted in knowing that he is always near and that we shared an awareness of the birds around us when he was alive. *Thank you, grandson.*

A cold gust of air and a reverent smile bring me back to the present. My grandson's sister, Rhea—which happens to be the name of a large ostrich-like bird found in South America—tosses her backpack and hat into the back seat of the car, then climbs inside.

Early-morning fog hovers above the ground and parking lot in front of the apartment. It's very early, even by birding standards, and her willingness—dare I say eagerness—to accompany me on yet another bird outing is both admirable and so very loving. Even more so given that she is a teen wrangling school and homework, extracurricular interests, household chores, friends, and social media, not to mention the grip of insomnia's nightly clutches. She has been my frequent birding buddy on numerous outings since she was a preteen and has attended a number of nature-based presentations and bird festivals. Therefore, she has mastered the art of being prepared to go, her water and maps at the ready, a bona fide birdwatcher and birder-in-training.

She and I know that the time spent together discovering new birds and greeting familiar ones is precious and special to just us two. These experiences have given us sacred memories. As a parent and grandparent, I have discovered a unique way to bond with my children and to set examples for future generations just as my parents did for me in their own way.

My mother, Fran, would point out to us kids the mourning doves as we drove around town and never failed to emphasize that they were a sign of peace and goodwill. I guess you can say that the tradition continues. Even when my grandchildren were mere infants, not yet able to walk, I pointed out the hummingbirds who frequented the feeders, and they were able to watch them with great delight as the birds zipped around. *Thank you, grandchildren.*

My mother was studying parapsychology in the early '70s and was an early participant of Synanon, an addiction treatment turned psycho-spiritual program that used encounter groups as a means for transformative and personal change. She was righteously ahead of her time in every way. She taught me to see what the Universe had to say through its many signs and symbols. All too often she was filled with fret, and inevitably peace and assurance would be bestowed upon her once she saw the gentle dove.

I came to associate all birds with good tidings. I knew from tangible proof that birds possessed the capacity for powerful healing because they were everywhere, and with them came peace for the observant. In my mother's recent frail years, I have been able to share her knowing with her again. Her brown eyes rimmed in an ocean blue from cataract surgery follow my fingers as I point out the jerky flight of house finches, black phoebes, and her mourning doves in the courtyard of her healthcare center. Her index finger will suddenly raise toward the sky along with her upward gaze, and still she shows me winged blessings. *Thank you, Mom.*

I have yet to go on a planned bird outing with any of my closest friends. *But I'm not you,* they say. *I like birds, but not the way you do,* they say. *I'm not sure if I can. How early in the morning did you say?* and, *How long do you plan to be out there?* they ask. *Let me get back to you,* they promise. We've known one another for a very long time beyond our marriages and long-term relationships, births, guardianships, and deaths, and life's many challenges and many more celebrations. And so I know they don't want to disappoint me; at the same time, they have no intentions—the key word being *intentions*—of going out to specifically watch birds.

No matter how many times I ask, coerce, or highlight the benefits, they ain't going. I have no plans of ever giving up the hope that someday they will change their minds though, and if you happen upon similar hesitance, neither should you. Their lack of interest in birds encourages me to point one out when we are out and about, and they cup their hand above their brow to shield the sun for a better view.

I am encouraged by these same friends, and some family members too, who live in different parts of the country and have different careers and lives. All will reach out with a "Hey, what kind of bird is this?" question with an accompanying photo that they've gone to great effort to take and share. I am encouraged by these wonderful souls who have done so while on travel from as far as Egypt along the Nile, while hiking a desert trail, from a game preserve in East Africa, and from the gardens and shores of Hawai'i—or who ask about some bird they've seen in some news article or happen to hear mentioned in conversation.

I am encouraged by the bird feeders they hang and the dishes of water they replenish each evening in their yards. I am encouraged by seeing them set back in place toppled nests with such care and the various birds they have come to routinely notice. And I am encouraged by their willingness to listen to my far-too-long descriptions of birds that I've seen and facts about a particular species, reinforced by their thoughtfully curated cards, blankets, jewelry, mugs, art, and books gifted to me along the way. *Thank you, family and friends.*

Friends, like family, need acceptance of who they are and what they find important to them. Honor your differences as much as your similarities, and accept each one without expectations in the spirit of love and compassion. Just as there are different types of feathers on a bird that vary for their respective function, our relationships are also diverse and varied and strengthened by our connections. It's this legacy of knowing and of sharing the reverence for nature and birds that will inspire us to continue to share the joy and wonderment of them with family and friends of all ages.

No age is too young or too old to begin. Appreciation for birds comes in many forms. For some it may be joining you or your birding group on an outing or pouring over a field guide together to try to decipher a mystery bird's name. For others it might be sharing photos from around where you live, work, and travel, and chatting about sightings you have independent of one another.

There are those whose special talents and inclinations lend themselves to gentle nurturance. There are also kind and observant souls who recognize the needs we have for a special kind of bonding and so acquiesce to being that one who willingly does so. Birdwatching is a wonderful opportunity to create lasting bonds, which create special memories and the opportunity for new traditions. I, for one, cannot think of a better way to celebrate a holiday, let alone an ordinary day, than to spend time with loved ones watching birds and strengthening our unique connections with one another.

Perhaps you've had the chanced blessing to come upon a feather in your path or have been gifted one from a loved one. In that moment of discovery, you probably admired its color, shape, and form and wondered how it came to be free from its bird and in your hand. You marveled at its construction as you ran your fingers gently along its light form, understanding the sturdy yet fragile nature of it and wondered about the bird's identity and its meaning to and for you personally. If a gift from another, you shared in the acknowledgment of the connection it brought you both. Such is the way of the feather. Such is the way of our relationships.

Unique only to birds are their feathers. The modern feather, that we admire so, is the evolution from the earliest theropod dinosaurs, which predated birds. It is now known that they form from feather follicles comprised of tubular organs distributed throughout the bird's skin. They grow into beautiful plumages of color, each along its central vane. Paleontologists and ornithologists have confirmed that with the eventual diversification of birds came adaptation of the feather with different sizes, shapes, and features serving different functions, including providing flight capabilities, protection, sensory monitoring, courtship displays, visual acuity, and species and age differentiation.[1]

The symbolic parallels between feathers and human relationships are striking. Our own origins reach back to time and place immemorial. The modern family is varied, made up of any combination of persons who share witness to experiences and memories that transcend the function of early humans, yet still a steadfast foundation is present, much like the feather's central vane, of what it means to be a family, a member, and its purposes. No two families are exactly alike, because its members are each unique with differences of age, personality, and purpose. We too are fortified by such differences.

Relationships tend to function wholly better in enriching and supportive ways when there is consistent care and tending by its members through maintaining contact physically, emotionally, and spiritually—that is, by maintaining and nurturing beneficial connections. You can share in the celebration of the glorious displays of birds' plumage with your family members and friends. The feathers' spectacular array of colors exceeds the imagination, and their iridescence and distribution can captivate our curiosity, such as the vibrantly hued neck feathers of a hummingbird or the plumed body of a starling when light reflects it just right. By sharing your experiences and extending invitations to join you with the intention for meaningful memories, you reinforce the special bonds with those you cherish and also create the space to form new ones.

Perhaps you have grown distant and have encountered difficulties communicating or relating with one another. By spending time with birds in a manner suitable for everyone, you can do an activity that can be enjoyed together free of tensions or disturbances. Also take notice and cherish those so dear to you who have transitioned, noticing the comforting messages brought to you by way of their winged spirit, their bird.

FROM INSIDE LOOKING OUT

You can employ the same principles for spending time with birds that you do for yourself with your companion. Just be sure to consider their needs and level of interest, including their age, physical and mental health, and the environment in doing so.

Initially, it may be beneficial for you to watch and listen for birds from inside where you are with your companion due to the comfort of familiar surroundings—your home or theirs, an office or building space, hospital room, classroom, or even from within a car.

Introduce the steps outlined in Chapter 1 to your companion. Either share this book with them or reference it on your own. It can serve as a guide to facilitate an introduction to the idea of birds and the power for them to heal.

Serve as a role model. Actively look for birds from where you are inside. If possible, position yourself by a window or doorway in the same area as your companion.

Share aloud what you hear and see. Gently indicate what you see in a manner that invites curiosity and the possibility for your companion to join you in a casual manner. Remember, even the act of listening is an active form of engagement.

Consider the length of time. Initially, it is probably ideal to keep the birding experience brief to prevent your companion from becoming overwhelmed or tired.

Enjoy the experience of spending special time with a loved one and don't dwell on the effort. While I recommend that you plan ahead for the most casual of encounters of shared birdwatching, it is equally important that you release expectations of the outcome. You and your family member or friend will deepen your connection through your genuine acceptance of who they are and what they decide to ultimately do. If they decline, graciously acknowledge what they express about your invitation. Lovingly express your appreciation for your time together and continue to try, try again on other days and times with an open heart and mind.

GOING OUTSIDE WITH ANOTHER

As with spending time watching birds from inside, when going outside to birdwatch refer to Chapter 1, Birdwatching at Home, and Chapter 2, Birdwatching in Your Community. The same principles you have learned can be used alongside your family member or friend.

Introduce birdwatching practices to your companion. A general overview of key points is a great introduction.

Take into consideration the needs and capabilities of your companion and select a setting that will offer them the best opportunity for enjoyment and success. Starting close to home or another familiar setting is ideal.

Consider the total time spent. It's important for you to factor in travel time in addition to the time spent outdoors. It is strongly recommended that you keep the first few outings to a short duration so that you both have the opportunity to evaluate the benefits and potential needs for future ventures together.

Share what you do know and don't know. By sharing your equipment, field guides, and notes, you help to extend compassionate support to your loved one. They will surely appreciate it and no doubt find courage in the process. It is a great opportunity for you to bond as you learn together.

Enjoy your exploration outdoors with one another!

STARTER TOOL KIT

The most important tool you will need is *the gift of your time and a well-planned invitation* to go birdwatching with someone you hold dear. Assist them with gathering the equipment and tools they may need. Initially, consider sharing your binoculars, scopes, field guides, food, water, transportation, and other equipment.

Include items from the Starter Tool Kits in Chapter 1, Birdwatching at Home, Chapter 2, Birdwatching in Your Community, and Chapter 4, Birdwatching and Your Health.

Family member(s) and/or friends
Preview your bird outing online and on-site.

(Optional)
Chairs or other supportive device for rest
Blankets

EXERCISE PROMPTS

Date/Time:

Spending time with a loved one watching birds is restorative and enjoyable. You can complete this exercise together or each on your own at your discretion.

Today, I am with:

We are:

inside outside

We are at *(enter your location)*:

This time, we choose to

sit stand lie down

and observe the birds.

First, listen for birds. We hear:

While you scan the area with your eyes, looking for signs of birds, take a slow deep breath in, hold it for four seconds, exhale slowly to a count of five, repeat.

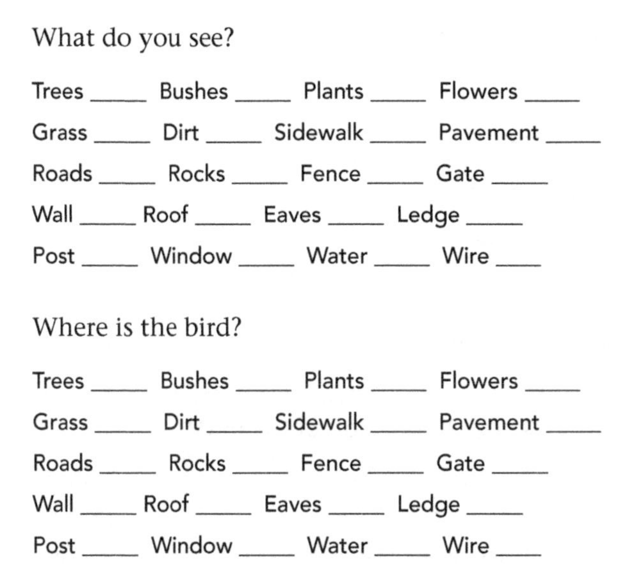

What do you see?

Trees _____ Bushes _____ Plants _____ Flowers _____
Grass _____ Dirt _____ Sidewalk _____ Pavement _____
Roads _____ Rocks _____ Fence _____ Gate _____
Wall _____ Roof _____ Eaves _____ Ledge _____
Post _____ Window _____ Water _____ Wire ____

Where is the bird?

Trees _____ Bushes _____ Plants _____ Flowers _____
Grass _____ Dirt _____ Sidewalk _____ Pavement _____
Roads _____ Rocks _____ Fence _____ Gate _____
Wall _____ Roof _____ Eaves _____ Ledge _____
Post _____ Window _____ Water _____ Wire ____

Take at least one full minute to watch the bird.

What is the bird doing? (*For example, sitting on a branch, walking through the grass.*)

Why do you think the bird is doing that?

Describe the bird you are observing.

- Size
- Shape
- Color(s)
- Eye color
- Leg color
- Beak color/shape

What is your impression of the bird?

Relationships are like feathers on a bird—they are interconnected in diverse ways and help the bird to thrive.

My family is:

As a family member, I am:

As a friend, I am:
Indicate specific relationships with someone close to you (daughter, best friend, grandson, co-worker, etc.):

My relationship with my ______________ (relationship), ______________ (name), is:

and we are:

Reflect upon a time a loved one who has transitioned reached out to you through a special bird.

What is the bird?

What was the message and its meaning for you?

How has your experience deepened your connection with your loved one?

I admire the feathers of the (*name of bird/description*) Why?:

Reflection Pond

We cannot see our reflection in running water.
It is only in still water that we can see.

—Taoist Proverb

Answer the questions as though you are a crane standing at the water's edge, gazing into a still pond at your own reflection. Invite your family member or friend to contemplate their own Reflection Pond.

Before I went birdwatching today with ____________,
I was:

While we were birdwatching together, I felt:

And now, I feel:

CHAPTER SIX

KEEP LOOKING UP

Birdwatching and Your Mental Health

No matter the day, always they beckon the morn. It begins when all is still and quiet and you can listen and heal.

Lying here, with eyes still closed, I am awakened reliably by their gentle chirps, hasty zips, and lyrical chatter. I recognize most of what I am listening to because they share the space around where I live—the wrens, song sparrows, house finches, mourning doves, hummingbirds, and spotted towhees, and in certain times of the year, the hooded orioles.

I am called to begin my morning meditation and gratitudes just as the dawn chorus ascends with the arrival of the sun, the bird song fuller and enveloping all around. *Thank you for this that I am experiencing right now, thank you for all that I can do, thank you for life, and thank you for another day. I am so very grateful.*

I open my eyes and look out the windows that are smaller than most typical bedroom windows and situated higher up near the ceiling. The tipa tree limbs stretch way up high, up to where I am on the second floor. I imagine that I am in a tree house looking out to the morning sky with its special kind of blue cast that hovers between night and not-quite-full day. I am at the top of the canopy, and the tiny bushtits and warblers scamper so freely as they pass by.

It's a beautiful daily reminder to commune with my feathered friends as they begin their day—slow, gentle, melodic, and gradually building to a crescendo at the day's full dawning. I join this dawn chorus every day.

I begin my day feeling lighter, happier, and already more connected to nature's healing. I reflect on how connecting with birds has helped me to transcend my depression, anxiety, and grief and cannot imagine living without them in my life.

Birdwatching is a practice and therefore needs to be done with regularity in order to reap the most benefits, just as when you learn to play an instrument, a new sport, or learn to speak a language. This holds true as well for the most effective strategies in addressing and maintaining your mental health. Birdwatching to help cope with your symptoms can be adjusted and customized to fit your own needs, and there is truly no wrong or right way, which is liberating in and of itself.

The fact that I am a mental health practitioner did not inoculate me from being mired in grief and depression. I prided myself on my ability to observe human behavior and accurately assess the psychosocial needs of others and outline appropriate clinical treatments to help them. And in the midst of my truest truths, I was able to know that I was secretly suffering and yet wasn't able to fully actualize those same keen treatment modalities to help myself.

There can be a myriad of causes for someone to develop clinically determined mental health conditions. In my case, it was due to medically induced injuries resulting in debilitating chronic pain. Upon unharnessed reflection, I am able to trace my predisposition to depression as being planted in my youngest of childhood years and cultivated into young adulthood. By middle age, I was indeed a master gardener, adapt at weeding out and managing symptoms better, or so I thought. My anguish and struggles to not succumb to depression caused me to have more anxiety, and the shame at not being able to handle it all led to even more shame, isolation, and prolonged grief from great loss.

It wasn't until I was a much older adult, homebound with disabling conditions, that I was finally forced to face my complete predicament—and therefore, my complete self—unabashedly and straight on. Initially, I was caught up in the fight to get back to what I had long been doing—working long, long hours as proof of my significance and running from one errand to the next on days "off" with little time for meaningful pauses and rest, and in hindsight, missing opportunities for quality, relaxed free time, and fun with my family.

Once it seemed undeniable that I would not improve, I reluctantly succumbed to the belief that my impaired physical and mental functioning from worsening chronic pain were here to stay. In turn, this process of detachment unknowingly fueled my symptoms of depression, anxiety, and grief even more. And so the cycle goes: the longer you endure, the more you experience these symptoms, which in turn manifests even more dis-ease.

When I think about it, I did know as a young child that I was often secretly sad and down. The stress of each day presented itself before I awoke and existed like my breath or heartbeat, constant and predictable. A nameless friend who hung around even though they were unwanted, and still it remained, steadfast and true, by my side. For the longest time, I knew my depression as oppression, fear, and fatigue brought on from circumstances beyond my perceived control, save my school-related accomplishments, my choice of friends, my time spent in real and imaginary play—usually outdoors and among animals, especially with my pet duck.

On the one hand I was well cared for with plenty of food and clothing, better-than-adequate shelter, and enviable parental doting. And on the other hand, I was fiercely and intensely encouraged, pushed, and harshly disciplined physically and emotionally to optimally fulfill my supposed talents and gifts as assurance for a prosperous future. Refusal and failure were not an option for me, and so from a very young age, I learned to suppress my emotions and feelings. It seemed to me to be less stressful, safer, and ultimately more rewarding to be quiet and compliant.

Depression is anger turned inward. I was frequently very angry at circumstances that I was made to endure in rigid silence while anxious about whatever chaotic episode loomed next. At the same time, I found great joy when I excelled academically, was heralded by family members for jobs well done, and comforted in the knowing that my family was devoted to making sure I had a happy, loving childhood.

Knowing that those charged with providing care and nurturance cannot help tame their afflictions' unfurling is, and isn't, salvation particularly for a child. It is a breeding ground for generational transmission of unintended mental instability and an existential threat that usually manifests as chronic depression and anxiety without the specific skills or tools to properly cope.

Without interference, a child knows life's purest truths, including the genesis of mental suffering, for a good while before others' opinions or designs take hold. And even then, they haven't the filters to screen them. Adults can mentally cast back to their years as children and understand that what was around them was, in fact, not normal and customary. And they can ultimately come to the realization that there were folks who were deeply afflicted, taken over by troubles they thought healed long ago. Such experiences can still permeate our consciousness when certain circumstances and memories arise or are unveiled.

The mother, who was too often bed-bound with worry and strain and therefore unable to care for herself and others, the gregarious uncle whose visit was dreaded because he became mean and ornery after dinner and countless alcoholic drinks, and the sibling who was always sullen and sad. Perhaps it was you in earlier years. Perhaps it is you now.

I learned to cope with my underlying depression by suppressing it or, at other times, bringing it along and overriding its desire to fade away who I truly was. We bargained, depression and me, sadness and me, anxiety and me, grief from loss and me. It was mutually agreed upon that there was to be coexistence, and I would take the lead in false pretense when it really counted—at school, with friends, and around most family, and as I grew older, at work as well. And what I had so convincingly assured myself as having effective coping skills was really denying myself proper self-care.

According to both the National Institute of Mental Health (NIMH) and the Centers for Disease Control and Prevention (CDC), hundreds of millions of people suffer from depression, chronic pain, and anxiety, which results in devastating isolation and a reduced quality of life. Worldwide, 1.5 billion people struggle with symptoms of depression and anxiety.[1] Typically, we experience limited mobility, are frequently opioid dependent with subpar efficacy, and rate our quality of life as poor to very poor.[2]

The good news in all of this is: (1) truly you are not alone in what you are experiencing and (2) you can learn to disrupt this vicious cycle by integrating birds into your life. As we've discussed, you can engage with birds in different ways in order to reap the many benefits. Specific mental health conditions have been shown to respond favorably. Here we will explore three common ones: depression, anxiety, and grief and loss.

One's early beginnings can lead to a predisposition for depression, anxiety, and undertones of unhappiness and isolation just as much as being social, relaxed, and often filled with joy. Some can be attributed to what life hands you and your biology, and on the other hand and of equal if not greater influence, some can be attributed to environment and nurturance. Nature and nurture.

Ultimately, it's *how* we manage our thoughts, feelings, and responses to circumstances, conditions, and states of mind that can define and impact how we behave and respond emotionally. Coping with adversity and high notes in life are in part learned through an almost innate transmission from caregiver to child, and so on through generations. This can include the triumphs and vestiges of cultural, political, and societal imprinting and can make it difficult to separate oneself from such legacies. In other words, we learn early on from those around us how to cope with circumstances, which may not always serve us well.

With all of these conditions, we experience somatic symptoms, meaning that they are physically felt in your body, and psychological symptoms, from your mind's thoughts and through your emotions.

DEPRESSION

Major depression is clinically defined as having symptoms that last for a majority of each day, every day for at least two weeks, with significant impact resulting in persistent impairment in the areas of work, school, and various social interactions. Common somatic symptoms include increased fatigue and lack of energy, changes in appetite, difficulty with sleep (either too much or not enough), and indiscriminate body movements such as pacing, wringing of hands, slowed speech patterns and motor reflexes, and poor concentration.

Psychological symptoms, that is nonsomatic ones, can frequently include feelings of worthlessness and guilt, sadness, emptiness and hopelessness, anger, and irritability. Thoughts that interfere with decision-making, create lack of interest in activities previously enjoyed, and are about suicide and death* referred to as suicidal ideation, are also commonly experienced.[3]

Depression can be downright depressing to think about. It can take over our whole being and signal to our body and mind that all is not well. You feel actual measurable and insidious hard-to-pinpoint physical and mental pain. It is exhausting and it makes you beyond fatigued. The seemingly easiest of tasks are gargantuan and overwhelming to tackle, and curling up in a ball as though back in the womb's protective cocoon seems like the only option. You fret and fuss with yourself, telling yourself to do better, be better, and to just get up and go. It sounds so easy—it seems to be easy enough—and yet you cannot. But you, who you know yourself to be, is in there, present and desirous of release.

Spending time in nature and watching the birds can help you. Begin where you are. If you cannot go outdoors, find a spot from inside where you are able to look outside.

Listen and watch for the birds. Try to do this a little each day. Observe the birds a minute or two at first and build up gradually over time. There is no hurry, and small, small steps will still make their way along the path. A bird builds its nest by making many trips back and forth carrying a blade of grass, a few horse hairs, a spindle of web, or a twig. Miraculously and with consistent tiny steps, its young are safely nestled within. The key to experiencing a bit of reprieve is to take the time and grant yourself permission to proclaim recognition for each tiny step.

---

*If you are experiencing frequent thoughts about suicide and death, it is imperative that you seek help immediately. In the United States, contact the 988 Suicide & Crisis Lifeline: CALL or TEXT 988. For International Suicide Prevention Assistance: go to Suicide.org. All hotlines are accessible for you to speak with someone and to receive help 24 hours a day, every day, including holidays. You may choose to remain anonymous.

On days that you can, sit outside where the birds are and just watch them for a while. The act of doing this will lift your mood and you will become naturally relaxed. The act of watching them creates active connectedness to nature, and your body begins to release what has been held within.

ANXIETY

Anxiety is depression's cousin, and both can be experienced concurrently, either as major independent conditions or as specified features within each diagnosis. There are five major types of anxiety disorders: social anxiety and phobia disorder, panic disorder, post-traumatic stress disorder (PTSD), obsessive-compulsive disorder (OCD), and generalized anxiety disorder (GAD). We will be considering GAD for discussion, though suggestions to alleviate symptoms can be applied regardless of the type you may have.

Unlike depression, to be clinically diagnosed with generalized anxiety disorder, symptoms must last most days for at least six months. Its prevailing feature is excessive anxiety and worry[4] that results in significant interference in daily life, including work, school, and social interactions. This is important to understand, because GAD is distinctly different from the occasional fear or anxious response you may understandably have to a particularly unsettling circumstance or event that you inevitably encounter through the course of your life. It is commonly expected to have a natural response to stress, feeling overwhelmed, or threatened.

General anxiety permeates your thoughts about your health, relationships, and routine life occurrences. It is characterized by several symptoms, including somatically experiencing tensed-up muscles, rapid heartbeat and breathing with or without heavy perspiration, shaking and trembling, easily becoming fatigued, difficulties with falling or staying asleep, the inability to concentrate, and digestive issues. Emotionally and psychologically, symptoms include irritability, feelings of restlessness and constantly being on edge, a general sense of panic and impending doom even in the absence of such threat, and excessive, constant worry and anticipation of negative outcomes. For those of you who are suffering with anxiety, your thoughts and perceptions about situations are plagued by the anticipation of potential negative outcomes, which fuels your excessive and debilitating worry.

Just as depression can be depressing to think about, anxiety convincingly causes you to feel anxious about feeling anxious! And yes, it is excruciatingly debilitating and exhausting to have to worry about worrying so. Our bodies are designed to function most ideally when in balance. As with depression and other health-related conditions, it is your body's way of letting you know that you are imbalanced and need help to gain mental and physical alignment.

Despite anxiety's persistence, incorporating the intentional connectedness to birds in nature is a beneficial strategy that you can add to your resources to combat it.

The key to reducing your symptoms of anxiety is to notice the first warning signs being signaled by your body: the first gastric pang in your stomach, the early tightening of your throat, the negative thoughts that begin to cycle, the changes from your typical breath rhythm to an escalating rapid and more shallow one, and feeling tired yet being unable to settle down.

It is at this point that you can go outside and immerse yourself in natural settings. The greener, the better. The more birds, the better. You can walk, sit, run, wheel, play, hike, exercise, climb, or swim, for example. Whatever it is that will release the tension building up in your body and mind. As you engage in your chosen physical activity, look for and notice the birds. Mentally acknowledge them as you can—"I see you robin," "Hey, cardinal, lovin' your red today," "I see all of you seagulls over there!"

As with other techniques designed to reduce your symptoms, it will take practice. It's a good thing that we are fortunate to have birds around us all the time in one manner or another.

GRIEF AND LOSS

Erich Fromm, the German-born American social psychologist, psychoanalyst, and humanistic philosopher once wrote, "To spare oneself from grief at all costs can be achieved only at the price of total detachment, which excludes the ability to experience happiness."[5] That is to say, loss and grief occupy equal space alongside attainment and joy. Recognition and acknowledgment of one's losses gives presence and voice to what it is you are enduring, which is grief.

Most often, folks associate the loss of a loved one with their own and others' grieving or bereavement. And while this is true, it bears emphasizing that you can grieve the loss of not only loved ones, but also your pet who dies, a relationship that has dissolved, a job or profession that has ended, a goal or dream not realized, and possessions and property lost. Know that it is to be expected that such losses will bring about a host of profoundly felt emotions, including sadness, shock, disbelief, and bewilderment. Most folks transition through these various stages and symptoms of grief over time and are able to function satisfactorily despite their loss and feelings about it.

However, when symptoms of grief and loss significantly interfere with your ability to function for an extended period of time, which is referred to as complicated grief, you may notice that symptoms of major depression take hold. Experiencing grief is immeasurably painful, yet it's a natural and necessary part of your journey toward eventual healing. In order to heal, we must acknowledge grief's demands for our attention and our hearts.

All too often we can feel pressured by our own as well as others' expectations to just "get over it" and behave "normally." In an effort to caretake others' feelings, you deny yourself the self-care that you so desperately need and deserve. It is usually unintentional and well meaning; nonetheless, it interferes with your natural needs toward ultimately finding healing that is best for you.

Honor your loss by granting yourself and those around you permission to be so very sad and longing for what was, and to spend the time in sorrow. Why wouldn't you be sad and in pain? In the more-than-human world, several species of wildlife—crows, elephants, and whales for instance—grieve their losses and mourn for notable periods of time. And then, following these periods of reverence, in some cases in solitude, they eventually rejoin their flock and are able to continue on.

Spend time outdoors for as long as you can each day. If you cannot be outdoors, spend time from inside looking out. Watch for birds to fly by, chirp in the bush, and perch near you. Think of your loss with intentional thoughts. As the bird song stops and the birds fly off, allow your feathered friends to take just a bit of your grief with them on their wings. They can carry your load regardless of its size, as they are mighty in spirit and came into your presence just long enough to help you toward healing. Watch the birds go for as long as you possibly can. Look up and offer gratitude for your time spent with them.

KEEP LOOKING UP

In the world of birdwatching and birding, looking up is a near necessity in order to see and even listen to birds, which is why, in part, doing so informed this book's title. I often find myself leaning way back at the base of trees while looking up through the canopy's crisscross of branches in search of birds' flutters and artfully crafted nests. Time and time again, I repeat this endeavor and see feathered friends. I am not alone in this postural stance, as I have seen many others with binoculars and cameras up to their eyes also looking up.

My personal experience of looking up through my kitchen window in glorious wonderment of the yellow warbler connected me to not just the outdoors again, but most importantly, to my powerful healing from watching birds.

Birds are usually perched up high somewhere, typically in a tree, on a utility line, or on a rooftop. Spring ushers in the much-anticipated bird migration and with it, the hallmark birdwatcher's signs of warbler neck, so named for stiffened necks, tensed-up muscles, and achy shoulders and arms acquired while painstakingly and continuously looking upward to spot feathered friends for extended periods of time. Warblers, in particular, are known to be tricky to spot as they flit higher and higher through leaf cover, though they offer delightful rewards—mostly rich shades of yellows, blacks, whites, and occasionally some greens and blues—when finally seen.

FROM INSIDE LOOKING OUT

Grant yourself the gift of communion with the dawn chorus. Dedicate your first precious moments upon awakening, before you rise, to reflection and intention. Try to begin even before your eyes open and listen to the birds' songs as they float along the still-early morning air, just before sunrise. *What are you grateful for?* Just as the birds alert you to their presence at the start of their day, the act of giving voice to your gratitude while immersing yourself in the call and song of the birds will create a positive mindset for the start of your day as well.

Choose a window or doorway that offers you the best vantage point to see outside.

Sit comfortably and watch for birds. Have no agenda other than to spend some time with the birds. Remember, start slow and gradually increase your time. Small steps will take you along the path.

When you cannot look or be outside, watch a video about birds in the wild. There are all kinds of videos available through bird clubs and organizations, streaming services, and social media platforms that share birds in their natural habitats. Studies support the benefits of viewing birds virtually for reducing symptoms of depression and anxiety, among others.

Watch a bird feeder and/or water feature. Attracting the birds to spend additional time where you can view them on a regular basis can help to improve your mood.

Journal your thoughts while sitting and observing the birds. There is no right or wrong way to do this, and it is completely at your discretion. *What are you feeling? What are you not feeling?*

Sketch a bird that you take notice of the most. Remember, it doesn't have to be perfect; in fact, a rough sketch is all you really need. The act of drawing your chosen bird is another way for you to connect with them and nature. It is relaxing and enjoyable.

GOING OUTSIDE AT HOME AND BEYOND

Several studies have been able to reinforce what many have anecdotally understood for years: spending time outdoors around where you live, preferably in green spaces like parks and gardens, is directly correlated with improved mood, reduced symptoms of depression and anxiety, and an overall feeling of being happier. Birds in particular have been found to bring joy to those who see and hear them. In fact, in one study, the more birds there were, regardless of type, the more mental health improvements there were among residents.[6] Researchers have also been able to compare various types of environments that people live in such as urban, rural, city, and suburban. They determined that spaces abundant with more shrubs, trees, and birds had inhabitants that reported less symptoms of depression and stress.[7]

Incorporate the steps for going outside as outlined in Chapters 1 through 8.

Awaken with the dawn chorus with intention and gratitude. Formulate a plan to go outdoors at some point in your day if you can. Doing so will provide an intentional shift in your personal self-care that will improve your mood and motivation.

Join a group outing, when and if you are able. Research supports engagement in nature-based interactions is physically, mentally, and emotionally healthy. In an evaluation of the Walking for Health program in England, it was found that "controlling for other significant predictors, group walks in nature were significantly associated with lower depression."[8]

Read stories, nonfiction and fiction, that share the struggles of managing mental health issues and the triumphs experienced. Research supports a reduction of depression and anxiety symptoms from reading self-help books.[9]

Talk with a trusted friend, family member, or other person, including culturally specific support such as a religious/faith leader elders, a healer, or a guide. It can be extremely helpful to share what you are going through with those you can trust.

Participate in mental health treatment. There are a myriad of treatment modalities and approaches that are available, including individual and relational therapy, family therapy, group therapy, and support groups. Services can be obtained in person, via online formats commonly referred to as telehealth, and via phone and text.

STARTER TOOL KIT

Include items from Starter Tool Kits from Chapter 1, Birdwatching at Home, Chapter 2, Birdwatching in Your Community, Chapter 3, Birdwatching as Mindfulness Meditation, and Chapter 4, Birdwatching and Your Health.

(Optional, but recommended)
A birdwatching buddy, in-person and/or virtual

Journal/sketchbook and pencil

EXERCISE PROMPTS

The dawn chorus ushers in a new day.

Today, I am grateful for:

Choose a tree to watch that you can visit regularly, from inside or while outdoors. Repeat this exercise as often as you can.
Date/Time:

My tree is *[include location description, type (if known)]*:

From Inside Looking Out

When I look up at the tree, I observe *(describe the birds, other wildlife)*:

Going Outside

When I stand under it and look up, I see *(describe the birds, other wildlife)*:

When I sit beneath it and look up, I see *(describe the birds, other wildlife)*:

MENTAL HEALTH SELF-ASSESSMENT

Free of shame and blame, my healing begins.

You can use this Mental Health Self-Assessment on the next page in two different ways:

1. As a daily journal: Fill it out on a regular basis.

2. Fill it out twice: once *before* and once *soon after* you have spent time watching birds.

Mental Health Self-Assessment

Today's Date/Time:											
My current mental health is:											
Depression:											
Circle one:	0	1	2	3	4	5	6	7	8	9	10
		no symptoms			some symptoms				many symptoms		
My symptoms/feelings are:											
Anxiety:											
Circle one:	0	1	2	3	4	5	6	7	8	9	10
		no symptoms			some symptoms				many symptoms		
My symptoms/feelings are:											
Grief and Loss:											
Circle one:	0	1	2	3	4	5	6	7	8	9	10
		no symptoms			some symptoms				many symptoms		
My symptoms/feelings are:											
Other condition:											
Circle one:	0	1	2	3	4	5	6	7	8	9	10
		no symptoms			some symptoms				many symptoms		
My symptoms/feelings are:											

Reflection Pond

We cannot see our reflection in running water.
It is only in still water that we can see.

–Taoist Proverb

Answer the questions as though you are a crane standing at the water's edge gazing into a still pond at your own reflection.

Before I rose today, I was:

While I spent time watching birds, I felt:

And now, I feel:

CHAPTER SEVEN

MIGRATION

Birdwatching in the Environment

I cannot believe that I am here—and yet here I am! They are here too, these "clowns of the sea," as puffins are affectionately referred to by admirers, bringing their delightful circus act back to remembered land. Not just any land though; here, at Haystack Rock in Cannon Beach, Oregon, where they take up annual temporary residence atop the uppermost grassy slopes of the world's third largest intertidal monolith. I am struck by the deep hues of earth serving as a backdrop to it. Even the Pacific Ocean meets the shore in a placid honoring of their magnificent presence.

The sunrise is creating a true ombré of reds and oranges. It is very early morning and crowds have yet to gather around. I stop for a few moments to speak my gratitude and take some mindful inhalations, savoring the

salted air and listening to the calls and responses of all—the birds, the ocean, and the crisp wind.

For this special trip, I have the incredibly good fortune to see the nature and wildlife that abound here, and in particular, these alluring birds, under the expert leadership of my guide and now friend, Jesse, a naturalist who selflessly volunteers her time, knowledge, patience, and contagious enthusiasm for marine conservation in the Oregon region. Without her, I may have not actualized my dream to even see these iconic seabirds.

She and I are standing ankle deep in tidepool swirls near the base of the massive monolithic rock. Without the aid of binoculars, I can just make out the flaxen-colored plumes, swooping oddly down from black rounded heads as they peek out from their burrows. Their faces boast their most distinguishing feature of all, a prominent triangular-shaped, bright-reddish-orange-colored bill surrounded by a very white face mask punctuated by black eyes rimmed in red. But it is the tufted puffin's breeding plumage that I most admire. Then these medium-sized black birds fly high over our heads, out over the water, and awkwardly land carrying clutches of long, skinny silver fish drooping in beaks for their one and only puffling.

Puffins are so beloved by the residents here that a festival is given in their honor, and June 2021 was proclaimed "Tufted Puffin Month" to celebrate these special birds and to broaden awareness of their needs to survive by returning here and concerted conservation efforts continue on their behalf.

It feels glorious to be on this road trip alone, finally manifesting a childhood dream of mine seeded back when I first peered at the magical portraitures of the puffins in the glossy pages of a *National Geographic* at the library.

Their penguinesque appearances captivated me so, and I thought that someday and somehow, I would have to trek to the far shores of Iceland, Wales, Alaska, Russia, or even the Auckland Islands, if I was to ever see them. As I became an adult, I thought it a surmountable endeavor born of undeniable necessity to delight *with* these comical birds, which is best done during their breeding season on land from their otherwise mysterious life out at sea where they migrate each year. The hows and whens of making this trip rendered themselves dormant over the many decades, but I never forgot about it or thought of it as impossible.

And so when I had the need to travel to Washington State to drop off my son, who was forging his own migration by relocating for the summer to rest and help out at extended family's rural property, I began to peruse our proposed route to determine what wildlife refuges, state parks, sanctuaries, and ecological preserves were along the way. If you are going to travel the road anyway, why not stop along the path and revel for a few minutes in the unique natural spaces the earth has gifted to us, along with the birds that inhabit them? It's a practice I soon learned to always incorporate in my travels near and far, despite even seemingly reasonable rationalizations to the contrary—expense, time, and added distance. Along the way, I have met many wonderful people, discovered hidden gems that would have gone unnoticed, and have come to deepen my appreciation for the impassioned conservation and restoration work done all over the world on behalf of our feathered friends.

Everything that has happened and everything I have learned leading up to this point has made it possible for me to venture thousands of miles from home in pursuit of sharing time with special birds in new places I'd never been before. The tools and practices of watching, listening, and spending time with birds at home and out in my community, alone and with family and with new birding friends, has garnered in me the confidence to go forth into the unknown filled with joy and excitement. The beauty of travel is the discovery of the unexpected and common marvels. No amount of preparation will take away the surprise of such travel gems, nor will the wonderment of your experience be denied.

Migration, in its simplest form, is a process of considerable movement from one point in time and place to another destination in time and place. Many species migrate across the continents in search of resources and improved opportunities, community, and safety, all driven by innate biological and environmental mechanisms.

Bird migration is one of the inexplicable phenomena of the natural world. Much is still unknown about the extent of biological and environmental forces that determine when and where species of birds take flight for other lands. Some birds, such as the bar-tailed godwit of New Zealand and the Arctic tern of Greenland, have been traced traversing some of the longest distances among bird species. They travel from one continent to the next to their respective breeding grounds by way of Asia, Europe, and South America. Birds go from ancestral breeding grounds to other places, then back again, which researchers aptly refer to as *philopatry* from ancient Greek to mean "loving fatherland."[1] Some, like the ever-present mallard ducks and northern cardinals, for example, never leave. The most common routes of travel for billions of birds, known as flyways, are between northern regions in the spring and southern geographical points in the winter, many of which breed in favorable northern habitats that offer plentiful food resources for short seasons, less competition, and better weather in which to rear their young.

We humans migrate for the same reasons as birds but are unique in also migrating to near and distant lands by pure choice of decision, for adventure, or a desire for change of scenery—or for all of these reasons. For some, returning annually to places we love full of tradition and in honoring of family and ancestors are highlights in life. We also have the same need for rejuvenation in familiar, comforting places away from our day-to-day bustle and for pure exhilaration. For many others, migration is driven by a robust need to provide abundant resources for ourselves and those we cherish, or for new explorations and discoveries.

Entire species of hawks and other raptors travel annually from South America up north and from Europe to the Middle East. You and your ancestors may be among whole populations who traveled just as far in pursuit of safety and happiness as well.

Travel opens our hearts and minds to that which is unfamiliar as much as that which is familiar to us. It is this convergence of experiences that offers us an opportunity to appreciate who we are, where we come from, and our own personal context among all else in the world. That is to say, you become more aware of your precious belonging in the Universe.

Your nest is your place at home. Your need to venture beyond where you live, work, and commune in regularity is, indeed, very strong. Listen to it and honor it accordingly, just as the migrating bird does. Circumvent the tendency to pay homage to your internal voice bellowing forth fear and trepidations, and the many reasons why you cannot and should not. It is common to experience such hesitancy and to be afraid of the unknown and the unpredictable, especially when traversing new lands. Quell such anxiety by, instead, reinterpreting your stifling internal dialogue as emotional and physical feelings of excitement and positive anticipation for what you *will* discover. By creating this new narrative about exploring beyond your nest and community, you will give rise to your need and desire to see more, do more, and fulfill more.

You can create space for a heightened spiritual connection along your journey and at your destinations by incorporating the practice of birding with intention as well as by happenstance. Both are equally rewarding in their own way. Watching and listening to birds, near and far, is a freeing experience and symbolizes the many journeys we take in life, both planned and not. Initially, practice birdwatching while traveling away from where you live yet still near enough in order to acclimate to being in new surroundings. Over time, you will develop an increased sense of independence and confidence that will support you going farther and discovering more about yourself and new species of birds.

The key strategy to a successful birdwatching outing in your environment while farther from home is unbridled curiosity, an open mind, and a loving heart. Pause and take notice of the beautiful birds that adorn your travel's paths. You will have rich and memorable experiences that you may otherwise not have endeavored to receive. You will return home filled with a renewed sense of "all is right with the world." And you will have a deeper connection with where you live, connecting it with the significance of birds to the health of our planet.

Enrich your entire being by witnessing the birds along your travels.

FROM INSIDE LOOKING OUT

Take some time and watch for birds from inside where you are. This includes from inside your lodging during your travels and from inside a vehicle, boat, train, bus, rickshaw, shuttle, or other forms of transportation you may use.

You have acquired the necessary skills to venture out birdwatching while exploring places away from home that may or may not be familiar to you. Perhaps you are returning to a favorite place, and this time you can experience it with a fresh perspective that incorporates birds and their habitats, which will deepen your connection and appreciation for the place and all that it has to offer.

Refer to the birdwatching practices outlined in the preceding chapters and summarized here:

Sit or stand as still as possible while watching.

Choose a window or opening that offers you the best vantage point to see outside.

Listen first.

Use your birdwatching observation skills to scan the area you are watching.

Bring binoculars and other related equipment as you deem fit for your location.

Use a field guide and apps. It is strongly recommended that you try to obtain field guides and related information about birds that are specific to the region you are visiting to enhance your ability to identify what you see, which will broaden your knowledge and overall enjoyment.

Take photos. Attempt to capture images of not only the birds you see but also the surrounding areas that you find them in. This will allow you to have enriched memories of your experiences seeing the birds and puts them into proper context within the overall environment for later when you go back through them.

Keep a birdwatching travel diary. Just as you are encouraged to keep a log of your sightings where you live, work, and recreate, also maintaining one while traveling will help you to chronicle your birdwatching journey. I recommend a dedicated diary, in book form and/or a digital one, that allows you to incorporate your sightings along with your travel experiences, such as the places you visit, the foods you eat, the activities you participate in, and the people you meet. Over time, you will have created a treasured record of a variety of your bird sightings specifically from your travels for your reflection for years to come.

Include sketches or a sketch book. Just as you have hopefully endeavored to sketch birds, other wildlife, and their habitats at home and in your community, you can also chronicle your travels in your sketch book. Consider including people, foods, and cultural and ethnic aspects that are not your own. Doing so will deepen your experiences and create a keepsake journal to depict your journey long after you have returned home.

Share your birdwatching experiences with your companions. At your discretion, point out the birds you observe to those you may be traveling with. In doing so, you introduce them to the practice of birdwatching from inside in a low-stress manner, free of demands and expectations. It also strengthens your travel and birdwatching experiences together.

GOING OUTSIDE IN THE ENVIRONMENT

Whether a short distance or thousands of miles away from where you live, take some time and develop a plan as to how you wish to consider incorporating your practice of birdwatching. As you have discovered on other kinds of birding ventures, pre-planning helps you have a more enjoyable experience by reducing the potential for missteps and increasing your safety.

Refer to the birdwatching practices outlined in the preceding chapters and summarized here.

Before You Go

Research the general region and specific location you plan to visit. Just as you might do when planning to travel to any destination, expand your research to include regions and areas designated for birdwatching and wildlife viewing, such as parks and forests, nature preserves, ecological reserves, sanctuaries, waterways and oceans, farms, and nature centers. Based upon what you learn as well as what piques your interest, plan your itinerary accordingly.

Do some virtual birdwatching in your target region. Whether it's because you are unable to go outside in the environment or you would like to familiarize yourself with the birds and their habitats ahead of time, watching videos and birdwatching outings virtually brings faraway birds and lands near. You can watch them repeatedly to thoroughly learn regional bird species and enhance your actual birdwatching outing. You will sharpen your ability to accurately identify these birds, which will boost your confidence and willingness to go birding in increasingly distant places.

Refer to Chapter 2 for more tips on birdwatching virtually.

Consider your safety. It is always a good idea to inform others of your plans, particularly if you are alone. Refer to the region's guidelines on overall safety, via governmental and tourism entities, and adjust your plans accordingly.

This may mean to include one or more birding buddies, the support of designated and trained security personnel, functioning cell phone service and other communication devices, accessible maps for areas without phone or Internet services, a whistle, and an experienced regional guide to accompany you.

Consider securing the assistance of a birdwatching guide. There are many reputable options available for guides who vary in their levels of expertise, size, and function. In addition to birding tour companies and private individual tour guides, also explore what is offered by local birding groups through organizations, universities, and museums. (See Resources.)

FACTORS TO CONSIDER WHEN CHOOSING A GUIDE

Level of Experience: Take careful consideration of your level of birdwatching experience and familiarity with the locale, and most especially that of the guide. It is strongly advisable to use resources that ensure you receive the assistance of well-informed birdwatching guides who are knowledgeable of your target destinations and the birds that inhabit them. Professional and skilled guides are typically accustomed to accompanying the novice as well as the veteran birder to the best locations and during optimal times in order to maximize seeing an abundance of endemic and even some rare bird sightings. An added bonus is if the guide is local and has knowledge of other wildlife, the culture of the people who live in the region, and opportunities for socio-cultural and conservation immersion, should you so desire to engage.

Guide and Tour Operations: Review the guide's proposed itineraries to determine the distance, duration, and terrain requirements. Inquire about what a typical day entails to ensure your expectations and needs are considered, including the proposed time to start and finish (will you leave early or late morning and then return evening or late at night?), opportunities for rest and down time, modes of transportation and trek, and engagement level with others who may be part of the organized group.

Cost: It is recommended that, when possible, you preview at least three different birding-guide services in the region you wish to explore and select the one that best meets your needs and is within your budget. Factors that contribute to cost include the size of the birding group, duration, locations, possible ancillary activities of the trip, travel and tour insurance, and personal preferences. For example, some prefer to have an individualized tour with just themselves and the guide and therefore are willing to incur the additional fees.

At Your Birding Location

Refer to and incorporate this chapter's steps from the section From Inside Looking Out *and from the preceding chapters.*

Review and/or compare regional and trail maps, guides, and apps that are posted, if available.

Bring assistive devices and equipment.

Factor in your physical and mental health needs and those of your companions, if known. *Refer to Chapter 4 and Chapter 6.*

Wear appropriate supportive clothing, footwear, and hat. Ensure that you have available the proper clothing and gear for the region you are in. This may give you the great opportunity to integrate culturally specific materials and tools as well, thereby supporting local economies and possible conservation efforts.

Preview weather conditions again and adjust plans accordingly. It is common for birders to go out birdwatching in mild-to-moderate conditions based upon personal stamina. You will still see birds in a bit of rain, for example, as they forage more so for insects and water.

Take your time and enjoy being where you are, watching and listening to the birds. Honor the time you have by yourself or with companions. Your experiences birding will never be the exact same twice. Savor the space you are in with acknowledgment and gratitude. Connect wholly and rejuvenate.

STARTER TOOL KIT

Taking the time to plan ahead is key to a successful birdwatching outing in lands beyond your home and community. This is particularly useful when you venture out into unfamiliar territory and will maximize your safety and enjoyment. You now have the necessary knowledge, skills, and insights to enjoy practicing birdwatching during your travels.

Include tools and strategies from Starter Tool Kits in Chapters 1 through 6.

Preview the region's birds, their habitats, and the surrounding communities.

Practice birdwatching virtually and learn about your targeted travel destination.

Consider your safety needs.

Bring assistive devices to help navigate various terrains and conditions, as needed.

Consider the assistance of a knowledgeable guide and/or birding tour company.

EXERCISE PROMPTS

Exploring new lands and the birds that inhabit them nourishes my soul.

Where is your preferred travel destination for rejuvenation?

Where have you dreamed of going? Why?

What steps can you take to make your dream trip a reality?

What birdwatching opportunities are there at or near your dream travel location?

Think about your most memorable trip.

What was your most memorable experience?

How did it enrich your life?

What birds did you observe?

Reflection Pond

We cannot see our reflection in running water.
It is only in still water that we can see.

—Taoist Proverb

Answer the questions as though you are a crane standing at the water's edge, gazing into a still pond at your own reflection.

Before traveling to ____________________, I was:

While birdwatching there, I felt:

And now, I feel:

CHAPTER EIGHT

RITUAL

Birdwatching for Life

Besides the city I live in, this house is the place I've lived in the longest in my life. I suppose it is the Old House now. It's where a lot of living has been done with the many ups and downs of it all. My children grew up here and attended the neighborhood schools with all the other kids who had moved into the new master-planned community, and so they were on equal footing and able to adjust well.

For a number of years there was a lot of home construction going on around us to establish nearby parks, fields, and trail systems to connect residents to the land and to one another. Small family-owned shops eventually

lined the main town square, centering around a grand fountain that would serve as the community's gathering space. In later years, grandchildren graced our home and lived nearby as well, making get-togethers commonplace.

Outside the house that we first moved into, there stands the same magnolia tree, a feature of a basic front-yard landscape with a typical suburban strip of grass put in by the builder. Every house in our neighborhood had the same kind, but for me, our tree had such special meaning. It signified potential, beauty, and the hope for more.

The backyard was hard-packed clay, inhospitable to even the hardiest of weeds, and cluttered with rocks of all sizes bordered by a fence. It was a blank canvas for our family, one the children found adventure in just as it was, playing fort, digging up and burying little treasures, and skinning knees. Eventually we had hardscape and landscape installed, complete with a lawn for the kids to play on, and for shrubs, plants, and flowers to line the edge along with young palms, jacarandas, and a tipa tree near the fountain.

I have lost count of the number of times I have sprayed glass cleaner on this kitchen window. It's a part of my daily ritual—a quick spray and cleaning of yesterday's dried water droplets from the kitchen sink in front so that I may have a clear view. I come downstairs and look out of it first thing each day, and following turning on the coffee to brew, I return once more and just look out and up. Way up, with my torso stretched over the sink so that I can see up, up, up into the gentle leaves of the tipa now equally matured as me. As it has grown skyward toward the sunshine, towering above the house's rooftop, I suppose that indeed it stands as testament to the passage of time for us all. I'm certainly grayer now, deeply middle-aged with smile and frown lines and flares of arthritis that seem to aggravate bulging varicose veins. The children are young adults now and live elsewhere.

I am so very thankful to be able to lean out and look up so far. Each time, hopeful to again meet the little yellow bird that came to me so long ago. I now know that the little bird that gifted me with such hope and connection beyond despair with its frequent visits back then was a feathered angel.

I haven't seen it or one like it since that first year. I now know that it was a male yellow warbler here during spring and its breeding season, singing his heart out for his beloved female somewhere near. He helped me to feel life's beauty and possibilities again.

Looking through this window over the many years is a revered ritual that exceeds being a mere habit, as each and every time I look up, I do so with intention and admiration. Others have taken up where the yellow warbler left off—the year-round residents of house finches, crows, hummingbirds, mourning doves, wrens, Cooper's hawks, goldfinches, and bushtits, with the bluebirds, woodpeckers, along with other kinds of warblers and vireos passing through on occasion. All come to the tipa tree, and many sip and bathe from the fountain's inviting flow directly underneath it.

My view out of this window is my most sacred space from inside where I live, looking out. Rituals that we have serve as acts informed by tradition, spirituality, culture, and society. They are behaviors practiced in a prescribed manner to convey meaning and reverence without necessarily spoken word. Symbols, signs, and beliefs combine with gatherings of one or more of us in designated sacred spaces in specified times—daily, annually, or when there is a life transition such as a rite of passage, death, or birth.

Birds have rituals as well. They display prescribed behaviors that are species specific and others that are universal. All are done to ensure that the avian world carries on. Most rituals that birds have are related to courtship and mating behavior, including elaborate presentations of colorful plumage, boisterous and melodic songs that are performed only under specific circumstances, complex nest construction and adornment, feeding from a male to a female to assure her of his caretaking skills for the young, and preening to invite safe proximity with light physical contact and closeness. Perhaps the most notable rituals observed are the carefully choreographed dances that move the birds across waters and land in such glorious synchronization.

These rituals are done in spaces and at designated periods of time set aside for territorial claim and the promise for nurturing young to fledge in full health, much like we do, at times. After all, rituals transcend generations and always help to propel life forward with significance and meaning.

Throughout this journey you have come to discover in this precious world the many avian wonders at home and where you live, work, play, and beyond. Along with acquiring new birdwatching skills, you have gained insight into your own processes for living your life, the opportunities for personal growth, and an understanding about your integral part with nature. Birding is a continual gift, with each day bestowing a unique experience for you to wholeheartedly embrace. You being you is good enough!

Birds and birding—and the joy they bring—is an opportunity to savor grace. There is no wrong or right way to connect with them. It can be just as you wish it to be. The birds will return the favor tenfold.

Create your own sacred space at home, another at a favored spot away from where you live, and as a member within the birding community. You can help to preserve the environments you live in, and in so doing, preserve the habitats for the birds to thrive. I continue on my journey toward good physical and mental health by practicing birdwatching on a daily basis in one form or another and in the manner that I am able, in full acceptance of who I am. I have become resilient through this journey to heal and am active in my community at the local, state, national, and international level through membership to environmental organizations, advocacy, and support.

The birds accept you unconditionally. Spend time with them indoors and outside, feed them, offer them fresh water, and commune with them and with those dear to you, whenever and wherever you are. Consider expanding your stewardship beyond where you live and advocate on birds' behalf through membership and volunteerism in the manner that best fits for you.

What you do in the world in support of birds and other wildlife is an opportunity for you to also provide yourself self-care. Remember, engaging in your community enriches lives, including your own, and contributes to your well-being.

I am excited and very encouraged by the earnest dedication to this necessary work. Such investments bode well for a more inclusive future that all of us can share as we better champion for the birds we cherish and for the habitats that they need.

It is my view that birds offer us important guidance about how best to live with and care for all of this planet's inhabitants. They show us to respectfully acknowledge and recognize our differences as welcomed strengths, understand that diversity begets stability, and that conscientious communities foster equal access to resources for everyone. You'll hear folks who work in conservation and the protection of the environment alongside birds say, "Birds do better when we do better."

Life will surely bring about some periods that are down and challenging, and many more that are uplifting, positive, and filled with joy. The beauty of birds is that they are our ever-present partners. That's the essence of the powerful healing a life with our feathered friends can give you, unconditionally.

May you forever remember to take time during each of your days to pause and look up for a while.

STARTER TOOL KIT

Fundamental to your toolkit is *your openness to integrating birds into your life,* ideally long term. You have traveled a full journey and have all that you will need to embark on your lifelong pursuits with birds.

Keep this book at a convenient place for easy reference, and take it with you on your birding journeys.

Consider designating a (sacred) space that offers you the opportunity to return to it regularly. From inside where you live, looking out, outdoors near where you live, and even farther.

Consider birds and birdwatching as integral parts of who you are and what it is that you do.

Commit to having and sharing the joy of birds and their powerful connections with others.

Have fun.

Bird for life!

EXERCISE PROMPTS

Coming full circle is to return to
sacred space, time and time again.

Complete this exercise using the same designated space. It can be filled out in two different ways:

1. Fill this out on a time interval of your choosing: daily, weekly, monthly.
2. Complete this for each season.

My sacred space *(from inside, looking out/outdoors)* where I live is:

I see *(include birds and wildlife)*:

What has transitioned/changed?

My sacred space beyond where I live is:

I see *(include birds and other wildlife)*:

What has transitioned/changed?

My favorite bird(s):

The bird I most identify my life with is:

I am grateful to birds for:

Birds I want to see in my lifetime:

Reflection Pond

We cannot see our reflection in running water.
It is only in still water that we can see.

–Taoist Proverb

Answer the questions as though you are a crane at the water's edge, gazing into a still pond at your own reflection.

Before I took notice of the birds around me, I was:

Now I notice birds and I feel:

For me, birds are:

RESOURCES

The world of birds is vast and inextricably interconnected with nature, the environment, and your well-being.

The following resources offer you various opportunities to assist with your journey to connect with our feathered friends and with all of the natural world for your mind, body, and spirit.

Remember to enjoy your process of discovery!

You can find additional resources at: tammahwatts.com

Bird Protection & Conservation Organizations

There are numerous environmental and conservation organizations throughout the world whose primary focus is birds and their habitats. Several offer comprehensive information, opportunities for volunteerism, and many resources.

National Audubon Society
www.audubon.org

American Birding Association
www.aba.org

American Bird Conservancy
abcbirds.org

Birdability
www.birdability.org

Birds Canada
www.birdscanada.org

Birding For All
birdingforall.com

Bird Life International
www.birdlife.org

Cornell Lab of Ornithology
www.birds.cornell.edu/home

Feminist Bird Club
www.feministbirdclub.org

The Mindful Birding Network
www.themindfulbirdingnetwork.com

The Royal Society for the Protection of Birds
www.rspb.org.uk

Nature & Outdoor Organizations

Over the years, the need for environmental and conservation organizations that take into account the considerations of our diverse communities has grown. They amplify the equitable access to nature for all.

Black AF in STEM Collective
www.blackafinstem.com

Children and Nature Network
www.childrenandnature.org

Disabled Hikers
disabledhikers.com

Latino Outdoors
latinooutdoors.org

Muslim Hikers-UK
muslimhikers.com

National Park Service
www.nps.gov/index.htm

The National Wildlife Federation
www.nwf.org

Native Women's Wilderness
www.nativewomenswilderness.org

The Nature Conservancy
www.nature.org/en-us

#NatureForAll
natureforall.global/home

Oceana
oceana.org

Outdoor Afro
outdoorafro.org

Outdoor Asian
www.outdoorasian.com

The Outdoorist Oath
www.outdooristoath.org

Queer Surf Club International
www.queersurfclub.com

Sierra Club
www.sierraclub.org

World Wildlife Fund
www.worldwildlife.org

Health

We all have health needs, whether to maintain a healthy and balanced lifestyle or to help address concerns and conditions. Consider exploring your health needs by searching name and/or category for specified information and support.

Alcoholics Anonymous (12-Step Addiction Support)
www.aa.org

American Chronic Pain Association
www.theacpa.org

Global Pain Initiative
www.globalpaininitiative.org

International Disability Alliance
www.internationaldisabilityalliance.org

International Pain Foundation
www.internationalpain.org

International Stress Management Association UK
isma.org.uk

Kripalu Center for Yoga and Health
kripalu.org

National Fibromyalgia Association
www.fmaware.org

Parks Rx America
parkrxamerica.org

Parks Rx Canada
www.parkprescriptions.ca

RSPB Nature Prescriptions UK
www.rspb.org.uk/natureprescriptions

SMART Recovery (non-12 Step Addiction Support)
www.smartrecovery.org

Spirit Rock Meditation Center
www.spiritrock.org

Mental Health

You are not alone. There are others who understand and are available to connect with you; most are available every day of the year.

988 Suicide & Crisis Lifeline
1-800-273-TALK (8255)
CALL or TEXT: 988

Crisis Text Line
Text HOME to 741741

American Association of Marriage & Family Therapists-AAMFT
aamft.org/Directories/Find_a_Therapist.aspx

Anxiety and Depression Association of America (ADAA)
adaa.org

Depression and Bipolar Support Alliance (DBSA)
www.dbsalliance.org

Good Therapy International Counsellors & Therapists
www.goodtherapy.org/therapists/countries

Grief Share International
www.griefshare.org

National Alliance on Mental Illness
www.nami.org

Psychology Today (Find a Therapist)
www.psychologytoday.com/us/therapists

United for Global Mental Health
unitedgmh.org

APPS

Apps available for your devices can assist with identifying birds, other animals, trees, and plants while on the go and where you live. Some offer you the opportunity to participate in community science by entering your observations and experiences to community, research, and scientific databases.

eBird
ebird.org/home

iNaturalist
www.inaturalist.org

Merlin
www.allaboutbirds.org/guide/merlin/id

ENDNOTES

Preface

1. Bret Stetka, "Bird Brains Are Far More Humanlike Than Once Thought," *Scientific American* (September 24, 2020). https://www.scientificamerican.com/article/bird-brains-are-far-more-humanlike-than-once-thought/.
2. Bret Steka, Bird Brains Are Far More Humanlike Than Once Thought, September 24, 2020, *Scientific American*.
3. Eldon Greij, "What Science Understands about Birds' Brains," BirdWatching, September 2, 2020, https://www.birdwatchingdaily.com/news/science/science-understands-birds-brains/.
4. Cornell Lab of Ornithology, *Handbook of Bird Biology*, ed. Irby J. Lovette and John W. Fitzpatrick, 3rd ed. (West Sussex, UK: Wiley, 2017).

Introduction

1. Zora Neale Hurston, *Dust Tracks on a Road: An Autobiography* (United States: McClelland & Stewart, 2019).

Chapter 2

1. Audubon, "The History of Audubon," Audubon, n.d., https://www.audubon.org/about/history-audubon-and-waterbird-conservation.https://www.audubon.org/about/history-audubon-and-waterbird-conservation.
2. Aududon, "This History of Aududon."
3. Cornell Lab of Ornithology, *Handbook of Bird Biology*, ed. Irby J. Lovette and John W. Fitzpatrick, 3rd ed. (West Sussex, UK: Wiley, 2017).

4. Gregory N. Bratman et al., "Nature and Mental Health: An Ecosystem Service Perspective," *Science Advances* 5, no. 7 (July 24, 2019), https://doi.org/10.1126/sciadv.aax0903.

5. Daniel T. Cox et al., "Doses of Neighborhood Nature: The Benefits for Mental Health of Living with Nature," *BioScience* 67, no. 2 (February 2017): pp. 147-155, https://doi.org/10.1093/biosci/biw173.https://theconversation.com/green-spaces-help-combat-loneliness-but-they-demand-investment-105260.

6. James K. Summers and Deborah N. Vivian, "Ecotherapy – a Forgotten Ecosystem Service: A Review," *Frontiers in Psychology* 9 (August 3, 2018), https://doi.org/10.3389/fpsyg.2018.01389.

7. "Mood Walks for Older Adults," Mood Walks, n.d., https://www.moodwalks.ca/mood-walks-older-adults/.

Chapter 3

1. Cornell Lab of Ornithology, *Handbook of Bird Biology*, ed. Irby J. Lovette and John W. Fitzpatrick, 3rd ed. (West Sussex, UK: Wiley, 2017): pp.162.

2. Swansea University. "Experts' high-flying study reveals secrets of soaring birds: New research has revealed when it comes to flying the largest of birds rely on air currents, not flapping to move around." ScienceDaily (accessed August 23, 2022). www.sciencedaily.com/releases/2020/07/200714111730.htm.

Chapter 4

1. Cornell Lab of Ornithology, *Handbook of Bird Biology*, ed. Irby J. Lovette and John W. Fitzpatrick, 3rd ed. (West Sussex, UK: Wiley, 2017).

2. K L Wolf, S Kreuger, and M A Rozance, "Stress, Wellness & Physiology: A Literature Review," Green Cities: Good Health (College of the Environment: University of Washington, 2014), https://depts.washington.edu/hhwb/Thm_StressPhysiology.html.

3. Jim Robbins, "Ecopsychology: How Immersion in Nature Benefits Your Health," *Yale Environment 360* (Yale School of the Environment, January 9, 2020), https://e360.yale.edu/features/ecopsychology-how-immersion-in-nature-benefits-your-health.

4. Mathew P. White et al., "A Prescription for 'Nature'—the Potential of Using Virtual Nature in Therapeutics," *Neuropsychiatric Disease and Treatment* 14 (November 8, 2018): pp. 3001-3013, https://doi.org/10.2147/ndt.s179038.

5. James Dahlhamer et al., "Prevalence of Chronic Pain and High-Impact Chronic Pain among Adults — United States, 2016," *Morbidity and Mortality Weekly Report* 67, no. 36 (September 14, 2018): pp. 1001-1006, https://doi.org/10.15585/mmwr.mm6736a2.
6. "Stress," National Center for Complementary and Integrative Health (U.S. Department of Health and Human Services, April 2022), https://www.nccih.nih.gov/health/stress.
7. "APA Dictionary of Psychology," American Psychological Association, n.d., https://dictionary.apa.org/stress.
8. "Stress," National Center for Complementary and Integrative Health (U.S. Department of Health and Human Services, April 2022), https://www.nccih.nih.gov/health/stress.
9. Melissa R. Marselle, Katherine N. Irvine, and Sara L. Warber, "Examining Group Walks in Nature and Multiple Aspects of Well-Being: A Large-Scale Study," *Ecopsychology* 6, no. 3 (September 19, 2014): pp. 134-147, https://doi.org/10.1089/eco.2014.0027.
10. K L Wolf, S Kreuger, and M A Rozance, "Stress, Wellness & Physiology: A Literature Review," Green Cities: Good Health (College of the Environment: University of Washington, 2014), https://depts.washington.edu/hhwb/Thm_StressPhysiology.html.
11. Daniel T. Cox et al., "Doses of Neighborhood Nature: The Benefits for Mental Health of Living with Nature," *BioScience* 67, no. 2 (February 2017): pp. 147-155, https://doi.org/10.1093/biosci/biw173.
12. MaryCarol R. Hunter, Brenda W. Gillespie, and Sophie Yu-Pu Chen, "Urban Nature Experiences Reduce Stress in the Context of Daily Life Based on Salivary Biomarkers," *Frontiers in Psychology* 10 (April 4, 2019), https://doi.org/10.3389/fpsyg.2019.00722.

Chapter 5

1. Cornell Lab of Ornithology, *Handbook of Bird Biology*, ed. Irby J. Lovette and John W. Fitzpatrick, 3rd ed. (West Sussex, UK: Wiley, 2017), pp 103-108.

Chapter 6

1. "Mental Health Information: Statistics," National Institute of Mental Health (U.S. Department of Health and Human Services, n.d.), https://www.nimh.nih.gov/health/statistics.

2. Haomiao Jia et al., "National and State Trends in Anxiety and Depression Severity Scores among Adults during the COVID-19 Pandemic — United States, 2020–2021," *Morbidity and Mortality Weekly Report* 70, no. 40 (October 8, 2021): pp. 1427-1432, https://doi.org/10.15585/mmwr.mm7040e3.

3. *Diagnostic and Statistical Manual of Mental Disorders*: DSM-5 (Arlington, VA: American Psychiatric Association, 2017).

4. *Diagnostic and Statistical Manual of Mental Disorders*: DSM-5 (Arlington, VA: American Psychiatric Association, 2017).

5. Erich Fromm, *Man for Himself: An Inquiry into the Psychology of Ethics* (Holt Paperbacks, 1990).

6. Daniel T. Cox et al., "Doses of Neighborhood Nature: The Benefits for Mental Health of Living with Nature," *BioScience* 67, no. 2 (February 2017): pp. 147-155, https://doi.org/10.1093/biosci/biw173.

7. Daniel T. Cox et al., "Doses of Neighborhood Nature: The Benefits for Mental Health of Living with Nature," *BioScience* 67, no. 2 (February 2017): pp. 147-155, https://doi.org/10.1093/biosci/biw173.

8. Melissa R. Marselle, Katherine N. Irvine, and Sara L. Warber, "Examining Group Walks in Nature and Multiple Aspects of Well-Being: A Large-Scale Study," *Ecopsychology* 6, no. 3 (September 19, 2014): pp. 134-147, https://doi.org/10.1089/eco.2014.0027.

9. "Books on Prescription: How Bibliotherapy Can Help Your Patients and Save Your Practice Time and Money" (National Association of Primary Care, September 2017), https://napc.co.uk/wp-content/uploads/2017/09/Reading-well.pdf.

Chapter 7

1. Cornell Lab of Ornithology, *Handbook of Bird Biology*, ed. Irby J. Lovette and John W. Fitzpatrick, 3rd ed. (West Sussex, UK: Wiley, 2017).

ACKNOWLEDGMENTS

It is with the deepest gratitude that I acknowledge all who have contributed to *Keep Looking Up: Your Guide to the Powerful Healing of Birdwatching* in their own special way.

I've said this many, many times, including in this book, and hopefully without sounding cliché: It takes a flock to arrive at a particular destination intact for what is to come. The one is strengthened by the power of the whole—a collective of souls.

I mean this from the depths of my knowing. This book is far beyond what I am or can do. It is a reflection of so many who have given their time, their energy, and indeed, it most certainly took a flock to bring this book to the world.

First and foremost, I am forever grateful to the Hay House team, without whom none of this would have been possible.

A very special thank you to Hay House President and CEO Reid Tracy for your inspirational words at the Hay House Writer's Workshop in Houston: "Write what you're passionate about." Along with birds, it changed my life profoundly.

Many, many thanks to my editor, Lisa Cheng, for your steadfast encouragement, expertise, and patient guidance. I also want to thank Monica O'Connor for your support, along with the rest of the Hay House team: Betsy Beier, Patty Gift, Celeste Johnson, Perry Crowe, Devon Glenn, Toisan Craigg, and the editorial team; Tricia Breidenthal and the design team, including Karim Garcia, Yvette Granados, Micah Kandros;

Lisa Reece; Kathleen Reed; Diane Hill and the sales team; Lizzi Marshall and the marketing team; and Lindsay McGinty and the publicity team.

I am forever grateful to Felice Laverne for your professional editing, superb coaching, and invaluable insight and to KN Literary Arts for providing foundational resources and knowledge in all things publishing.

To the special Hay House Authors Support Group (you know who you are!) and the Point Loma Group et al. (y'all know who you are too!), who have generously shared your perspectives which helped to solidify the book's title and vision—it is very much appreciated and remembered. To Sage Billick for your phenomenal styling and Dominique Labrecque for capturing me and the birds—your special connection to birds and their meaning made your gift of time to create the KLU vision a reality heartfelt.

To the entire Audubon California team and board, the Southern California Audubon Council, Buena Vista Audubon, San Diego Audubon, and the National Audubon Society teams—thank you for your birding guidance, support, consultation, and acceptance of who and what I am, including during a very tender point in my life, in every way, Bird Lady and all. Tom and Jesse, I cherish you and your loving guidance in nature.

Mike Lynes, thank you for seeing me and unknowingly reaching out within hours of me having submitted my book proposal to Hay House. It has forever changed the trajectory of my story with birds.

To Dr. Elizabeth Gray, Micah Mortali, Asha Frost, Sandra Hinojosa Ludwig, Kelly McDaniel, Holly Merker, Molly Adams, Sydney Golden Anderson, Abby Burd, Virginia Rose, and Dr. Ghada Osman—for your invaluable time and words, which lift up KLU far beyond what I could even imagine, I am eternally indebted to you. I have learned so much from you and the good work you do for so many around the world. Thank you so much for expressing my book's vision with such grace.

To the birding commUNITY—I am continuously grateful for your kindness and intentional amplification of KLU and all that it represents about our shared love for birds and living our truest lives. We have met along sidewalks, paths, and trails, on and off shore, in various parts of the world, both in person and virtually, and found in common our feathered friends.

And to my fellow souls who notice—and those who may not have noticed birds yet—I am grateful for your presence in the spirit of the bird, who knows no boundaries and defies limitations, and my hope is that you find peace each day alongside them. I certainly do knowing you are out there too.

To my Sister-Friends, I love you and am forever in deep gratitude for your precious time, lovingkindness, and being your awesome selves through it all.

Vekeno, for your warm Houston hospitality and serene home space, and the special time connecting, which fostered KLU's earliest and magical formation, your frequent reads, and both professional and personal advice along the way; Linda, for the frequent brainstorming of my ideas, which sharpened the message, your sage clinical advice, the numerous listening sessions in preparation for professional events in support of KLU, and acceptance; Audrey, for your celebration of our feathered friends and the beauty they bestow upon us that you share, which continuously inspires me despite being miles apart, for your early reads, and always-thoughtful reflections; Melanie, for showing up and being there; and Susan, for listening and cheering me onward with matter-of-fact conviction.

To my Family—you are my heart and soul, and I absolutely know that without your unconditional love and support, I could not have written this book.

My earliest memories are of me and my parents, James N. Helton, Sr., and Frances Helton, reveling in the outdoors in one manner or another—collecting shells, leaves, feathers, stones, and creating a life alongside birds, domesticated and in the wild.

Thank you dear children, bonus, and grands—Meghan, Morgan (Jon), Taylor (Taylor), Cameron, Sam (Angela), Alexis (John), Max, Rhea Buttercup, Gabe, Easton, Cy, Galileo, Michael, Esmerelda—for your deep appreciation for and connection with the more-than-human world, especially the birds all around. The essence of who you are as celebrated through your travels, art, music, writings, and pure joy and is the best gift I have received. You all shine so bright as wonderful examples of world peace and love.

I am also very grateful to my brother James (Laura) and my siblings, and to my extended family, including John Keith, Guy Anthony, Aunt Loneta, Uncle Calvin, Frank and Juliette Walen, Jeff and Christine Beaulieu and clan, Burcu, and the Watts family for your faith and encouragement that kept me going.

And to my husband, Harrison Watts—thank you, and I love you—it has been quite the writer's life journey indeed.

And finally, I must recognize my ancestors—those who have traversed the earth, many since time immemorial and passed along from one generation to the next a knowing about the land and all beings, including the meaning and importance of our feathered friends. I am them, and so in many regards, *Keep Looking Up* is a tribute to their very essence and therefore their stories.

ABOUT THE AUTHOR

Tammah Watts is a licensed marriage and family therapist (LMFT), certified Kripalu mindfulness outdoor guide and an advocate for the equitable accessibility of nature, health and mental health for all.

She has lived a life devoted to people and communities in need, having served for over 30 years in managerial, clinical and consulting roles for private non-profit, community college and public sectors.

Tammah has a MA in counselling psychology and is currently an associate mental health counsellor at MiraCosta Community College's Student Health Services, where she provides counselling and group therapy to a diverse student population.

She brings that same passion to conservation and birding and serves on the Audubon California Advisory board of directors as the Southern California Audubon Council Representative for 15 local chapters and college club, as well as Buena Vista Audubon board of directors on their conservation and equity, diversity and inclusion (EDI) committees.

Tammah offers keynote speaking, presentations and workshops and is co-host of the new Audubon CA interactive programme series, 'The Bird Story Hour,' which honours attendees' personal stories about feathered friends. She is inspired to share her story in hopes that the benefits of nature and the wonderful world of birds can resonate with others.

She continues to live in San Diego, California, with her husband Harrison and tries to combat empty nest syndrome by visiting with their adult children and grandchildren whenever possible. **www.tammahwatts.com**

Hay House Titles of Related Interest

YOU CAN HEAL YOUR LIFE, the movie,
starring Louise Hay & Friends
(available as an online streaming video)
www.hayhouse.com/louise-movie

THE SHIFT, the movie,
starring Dr. Wayne W. Dyer
(available as an online streaming video)
www.hayhouse.com/the-shift-movie

***

A FIERCE HEART: Finding Strength, Courage, and Wisdom in Any Moment, by Spring Washam

LET YOUR FEARS MAKE YOU FIERCE: How to Turn Common Obstacles into Seeds for Growth, by Koya Webb

TREE WISDOM: A Year of Healing Among the Trees, by Vincent Karche

THINK INDIGENOUS: Native American Spirituality for a Modern World, by Doug Good Feather

All of the above are available at www.hayhouse.co.uk

***